Signaling in ATM Networks

For further titles in the *Artech House Telecommunications Library,*
please turn to the back of this book.

Signaling in ATM Networks

Raif O. Onvural
Rao Cherukuri

Artech House
Boston • London

Library of Congress Cataloging-in-Publication Data
Onvural, Raif O., 1959–
    Signaling in ATM networks / Raif O. Onvural, Rao Cherukuri.
        p.   cm.—(Artech House telecommunications library)
    Includes bibliographical references and index.
    ISBN 0-89006-871-2 (alk. paper)
        1. Asynchronous transfer mode—Standards.   2. Signal theory (Telecommunication)
    I. Cherukuri, Rao.   II.   Series.
    TK5105.35.O55   1997
    621.39'81—dc21
                                                                                                97-30791
                                                                                                CIP

British Library Cataloguing in Publication Data
Onvural, Raif O., 1959–
    Signaling in ATM networks
    1. Asynchronous transfer mode   2. Signal processing
    I. Title   II. Cherukuri, Rao
    621.3'822

    ISBN 0-89006-871-2

Cover and text design by Darrell Judd

© 1997 ARTECH HOUSE, INC.
685 Canton Street
Norwood, MA 02062

International Standard Book Number: 0-89006-871-2
Library of Congress Catalog Card Number: 97-30791

10 9 8 7 6 5 4 3 2

For their love, Meliha, Nur, Melih, and Doruk
and of course Seval and Aytan

R.O.O.

With love for Pramela, Padma, and Mena

R.C.

Contents

	Preface	**xvii**
	Acknowledgments	**xix**
1	**Introduction**	**1**
1.1	Why Signaling?	3
1.2	ATM Interfaces	5
1.3	Signaling Structure in ATM Networks	7
1.3.1	The Signaling AAL	9
1.3.2	Primitives	13
1.4	Signaling Message Format	17
1.5	Information Elements	19
1.6	Characterization of an ATM Layer Connection	20
1.6.1	ATM Addresses	21
1.6.2	ATM Layer Traffic Parameters	23
1.6.3	ATM Layer QoS Parameters	24
1.6.4	ATM Transfer Capabilities	26
1.6.5	Multiconnection Call	28
1.7	ATM Standards	28

1.7.1   International Telecommunications Union   28

1.7.2   ATM Forum   36

1.7.3   American National Standards Institute   38

1.7.4   European Telecommunication Standards Institute   40

2   UNI Point-to-Point Signaling   41

2.1   Overview of Call/Connection Control   42

2.2   Information Elements   47

2.2.1   ATM Adaptation Layer Parameter Information
         Element   47

2.2.2   Broadband High-Layer Information IE   53

2.2.3   Broadband Low-Layer Information IE   54

2.2.4   Called Party Number Information Element   55

2.2.5   Called Party Subaddress Information Element   56

2.2.6   Calling Party Number Information Element   56

2.2.7   Calling Party Subaddress Information Element   57

2.2.8   Cause Information Element   57

2.2.9   Connection Identifier Information Element   58

2.2.10   End-to-End Transit Delay Information Element   59

2.2.11   Quality of Service Parameter Information Element   60

2.2.12   Broadband Repeat Indicator Information Element   61

2.2.13   Transit Network Selection Information Element   61

2.3   Call and Connection Control Messages   62

2.3.1   ALERTING   63

2.3.2   CALL PROCEEDING   63

2.3.3   CONNECT   64

2.3.4   CONNECT ACKNOWLEDGE   64

2.3.5   SETUP   64

2.3.6   Coding Rules for Information Elements   66

2.4   Call/Connection-Establishment Procedures   69

2.4.1   Connection Establishment at the Originating
Interface   69

2.4.2   Call/Connection Establishment at the Destination
Interface   74

2.4.3   Call/Connection Establishment—State Diagrams   78

2.4.4   ATM Adaptation Layer Parameters Indication in the
SETUP Message   78

2.4.5   Handling of the End-to-End Transit Delay
Information Element   82

2.5   Call/Connection Clearing   83

2.5.1   Call-Clearing Messages   83

2.5.2   Call-Clearing Procedures   85

2.5.3   Call/Connection Clearing—State Diagrams   87

2.5.4   Clear Collision   87

2.6   Restart Procedure   87

2.6.1   Messages and Information Elements   87

2.6.2   Sending a RESTART Message   90

2.6.3   Receiving a RESTART Message   91

2.7   Handling of Error Conditions   93

2.7.1   Protocol Discrimination Error   94

2.7.2   Message Too Short   94

2.7.3   Call Reference Error   94

2.7.4   Message Type or Message Sequence Errors   95

2.7.5   General Information-Element Errors   95

2.7.6   Mandatory Information-Element Errors   96

2.7.7   Nonmandatory Information-Element Errors   97

2.8   Signaling AAL Connection Reset   98

2.9   Status Enquiry   99

2.9.1   Messages and Information Elements   99

2.9.2   Status Enquiry Procedure   100

2.9.3   Receiving a STATUS Message   101

2.10   Error Procedures With Explicit Action Indication   103

2.10.1   Unexpected or Unrecognized Message Type   103

2.10.2   Information-Element Errors   104

2.10.3   Handling of Messages With Insufficient
          Information   105

2.11   List of Timers   105

2.12   Examples   105

2.12.1   LAN Emulation Over ATM   105

2.12.2   Video on Demand   110

2.12.3   IP Over ATM   113

3   UNI Point-to-Multipoint Signaling   117

3.1   An Overview of Point-to-Multipoint Signaling
          Framework   119

3.1.1   An Overview of Point-to-Multipoint Signaling
          Messages   119

3.1.2   Point-to-Multipoint Party States   121

3.2   Point-to-Multipoint Control Messages   122

3.2.1   ADD PARTY   122

3.2.2   ADD PARTY ACKNOWLEDGE   124

3.2.3   PARTY ALERTING   124

3.2.4   ADD PARTY REJECT   125

3.2.5   DROP PARTY   125

3.2.6   DROP PARTY ACKNOWLEDGE   125

3.3   Information Elements Used in Point-to-Multipoint
          Signaling Messages   126

3.3.1   Endpoint Reference Information Element   126

3.3.2   Endpoint State Information Element   127

3.4   Point-to-Multipoint Signaling Procedures   127

3.4.1   Connection Setup to the First Party   128

3.4.2   Adding a Party at the Originating Interface   129

3.4.3   Add Party Rejection   131

3.4.4   Add Party Establishment at the Destination
Interface   133

3.4.5   Party Dropping   135

3.5   Restart Procedure   139

3.6   Handling of Error Conditions   139

3.6.1   Endpoint Reference Information Element Related
Errors   140

3.6.2   Message Type or Message Sequence Errors   140

3.6.3   Mandatory Information Element Related Errors   141

3.6.4   Status Enquiry Procedure   142

3.7   Procedures for Interworking With Private
B-ISDNs   143

3.7.1   Incoming Add Party Request   144

3.7.2   Response to an Add Party Request   145

3.7.3   Party Dropping   146

3.8   Timers Used in Point-to-Multipoint Signaling   147

3.9   Handling the End-to-End Transit Delay
Information Element   147

3.9.1   Originating UNI (SETUP or ADD PARTY)   147

3.9.2   Destination UNI (SETUP or ADD PARTY)   147

3.9.3   Called User   147

3.9.4   Destination UNI (CONNECT or ADD PARTY
ACKNOWLEDGE)   150

3.9.5   Originating UNI (CONNECT or ADD PARTY
ACKNOWLEDGE)   150

3.10   Leaf-Initiated Join Capability   150

3.10.1   Leaf-Initiated Join Messages   151

3.10.2   Leaf-Initiated Join Information Elements   152

3.10.3   Root Creation of a Network Leaf-Initiated
Join Call   153

3.10.4   Leaf Joins to Active Network Leaf-Initiated
Join Call   154

3.10.5   Leaf Joins as the First Party to a Point-to-
Multipoint Call   156

3.10.6   Leaf Joins to Root Leaf-Initiated Join Call   158

3.10.7   Call/Connection and Party Clearing   159

3.10.8   Handling of Error Conditions   160

3.10.9   Timers at the User Side   160

**4   Signaling Support for ATM
Transfer Capabilities   161**

4.1   ATM Transfer Capabilities   165

4.1.1   Signaling of ATM Transfer Capabilities   168

4.2   Deterministic Bit Rate Transfer Capability   170

4.3   Statistical Bit Rate Transfer Capability   172

4.3.1   Signaling Procedures for the SBR Transfer
Capability   172

4.4   Available Bit Rate Transfer Capability   173

4.4.1   Signaling for ABR   176

4.4.2   Signaling Procedures   177

4.4.3   Allowable Combination of Parameters in the
SETUP Message   183

4.5   ATM Block Transfer Capability   183

4.5.1   Signaling for ABT Transfer Capability   184

4.5.2   Signaling Procedures   185

4.5.3   Allowable Combination of Parameters in the
SETUP Message   185

4.6   Frame Discard   185

4.7   Connection Characteristics Negotiation During
Establishment Phase   187

4.7.1   Information Elements   188

4.7.2   Negotiating the Connection Characteristics   188

4.7.3   Negotiation Acceptance   191

5	**B-ISDN Supplementary Services**	**193**
5.1	Direct Dialing In (DDI)	194
5.2	Multiple Subscriber Number (MSN)	195
5.3	Calling Line Identification Presentation (CLIP)	196
5.3.1	Actions at the Originating Switch	197
5.3.2	Actions at the Destination Switch	200
5.3.3	Interworking	202
5.3.4	Two-Calling Party Number Information Elements Delivery Option	203
5.4	Calling Line Identification Restriction (CLIR)	205
5.4.1	Interworking	205
5.5	Connected Line Identification Presentation (COLP)	206
5.5.1	Information Elements	207
5.5.2	Actions at the Originating Local Exchange	208
5.5.3	Actions at the Destination Switch	209
5.5.4	Interworking	211
5.6	Connected Line Identification Restriction (COLR)	214
5.6.1	Interworking	216
5.7	Subaddressing (SUB)	216
5.7.1	Normal Operation	218
5.7.2	Interworking	218
5.8	Relationship of Address Information Element and Supplementary Services	218
5.9	User-to-User Signaling (UUS)	219
5.9.1	Messages	220
5.9.2	Information Elements	220
5.9.3	UNI Signaling Procedures	221
5.9.4	Interworking	222

6	**Interworking**	**223**
6.1	ISDN Circuit Mode	224
6.1.1	Interworking Between N-ISDN and B-ISDN	228
6.1.2	Notification of Interworking	231
6.1.3	Tones and Announcements	231
6.2	Frame Relay Service	232
6.2.1	Mapping of Call Setup Messages	235
6.2.2	Traffic Parameters Mapping	236
6.2.3	Interworking	237
7	**Private NNI Signaling**	**239**
7.1	An Overview of the PNNI Framework	240
7.2	PNNI Signaling	249
7.2.1	An Overview of PNNI Signaling Messages	249
7.2.2	An Overview of Point-to-Point Call States	251
7.3	An Overview of PNNI Connection Setup	253
7.3.1	A Point-to-Point Call Setup	254
7.3.2	A Point-to-Multipoint Call Setup	256
7.4	PNNI Point-to-Point Call/Connection Control Procedures	257
7.4.1	Call/Connection Establishment	258
7.4.2	PNNI Point-to-Point Call Control Messages	263
7.5	Additional PNNI Signaling Procedures	268
7.5.1	Connection Identifier Allocation	268
7.5.2	Traffic Parameter Negotiation During Call/Connection Setup	272
7.5.3	QoS Parameter Selection Procedures	275
7.5.4	Designated Transit List	280
7.5.5	Crankback	287
7.5.6	Restart Procedure	295

7.5.7   Traffic Parameter Selection Procedures for ABR
        Connections   297

7.6   A Point-to-Multipoint Call Setup   298

7.6.1   First Party Setup   299

7.6.2   Adding a Party   300

7.6.3   Messages for ATM Point-to-Multipoint Call and
        Connection Control   301

7.6.4   Crankback Procedures for Point-to-Multipoint
        Calls/Connections   304

7.7   Soft Permanent Virtual Connections   305

7.7.1   PVPC/PVCC Information Elements   307

7.7.2   An Overview of the Procedures   309

8   Traffic Parameter Modification   311

8.1   Modification Signaling Messages   312

8.1.1   MODIFY REQUEST   312

8.1.2   MODIFY ACKNOWLEDGE   313

8.1.3   MODIFY REJECT   313

8.1.4   CONNECTION AVAILABLE   314

8.2   Additional Point-to-Point Call States   314

8.3   Modification Procedures at the Requesting Entity 314

8.3.1   Modification Acknowledgment   315

8.3.2   Indication of Modification Rejection   316

8.3.3   No Response to Modification Request   317

8.4   Modification Procedures at the Responding Entity 317

8.4.1   Modification Indication   317

8.4.2   Modification Acceptance   317

8.4.3   Modification Confirmation   317

8.4.4   Modification Rejection   318

8.5   Timers   319

About the Authors   **323**

Index   **325**

Preface

Broadband integrated services digital networks (B-ISDNs) are envisioned to be the universal communications framework, integrating all types of networking services and applications. Towards achieving this objective, a B-ISDN should not only provide a framework to support emerging multimedia applications but also support services that are currently deployed. Most corporations today use separate networks for voice and data services, with different protocols and different networking technologies. Emerging multimedia applications such as digital video, desktop videoconferencing, access to remote video libraries, and distance learning require service guarantees from the network. Current packet-switched networks are designed to support best-effort service only and cannot provide service guarantees. Circuit-switched networks can provide service guarantees, but they are inefficient in supporting data traffic.

Asynchronous transfer mode (ATM) networks are designed to support a variety of applications with different service requirements and traffic requirements. When a network supports applications with delay and loss requirements, the network must decide whether there are enough resources in the network to admit a new service request without affecting the commitments made to network applications already utilizing the network (i.e., admitted). Hence, a mechanism is needed for the end user to specify to the network its application's traffic characteristics and service requirements. Using this information, the network can decide whether or not it is possible to provide the requested service. In addition to the traffic-related information that needs to be exchanged between the end users and the network (or between two networks), end users may want to exchange some application-specific information (e.g., higher layer protocols) among themselves. Similarly, an end-to-end path between two or more communicating entities may cross one or more networks and/or one or more switches. The user-related information needs to be received and processed by these network elements in order to make the admission decision.

B-ISDN signaling provides the necessary information exchange between the end user and the network and between networks/switches. More formally, signaling in a communication network is the collection of procedures used to dynamically establish, maintain, and terminate connections. The signaling framework should be flexible to support additional capabilities as they are needed by the emerging multimedia applications. The most recent major milestone in B-ISDN signaling took place in late 1996, when the latest ATM Forum and International Telecommunications Union (ITU-T) specifications on user-to-network interface and private network-to-network (or network node when used between two switches) were finalized.

This book attempts to summarize the current status of ATM signaling. It is envisioned to be a one-source reference that explains in detail the current state of the art in signaling in a form understandable to an interested reader.

The book consists of eight chapters addressing different aspects of ATM signaling. Chapter 1 discusses what signaling is, presents a historical perspective, and quickly overviews ATM technology and standards organizations. Chapters 2 and 3 address user-to-network, point-to-point, and point-to-multipoint signaling protocols. These are based on ITU-T Recommendations Q.2931 and Q.2971 and include ATM Forum user-to-network interface version 4.0 changes and extensions, including leaf-initiated join, anycast capability, and NSAP (network service access point) ATM addressing formats. Chapter 4 presents a brief review of different ATM transfer capabilities: CBR (constant bit rate), RT-VBR (real-time variable bit rate), NRT-VBR (non-real-time variable bit rate), ABR (available bit rate), ABT (ATM block transfer), and UBR (unspecified bit rate). It also presents a detailed review of signaling procedures developed to enable these ATM transfer capabilities. Chapter 5 covers various supplementary services defined for B-ISDN and signaling to support these services. Chapter 6 is a review of the signaling support developed for interworking between ATM, frame relay, and integrated services digital network (ISDN) services. Chapter 7 addresses the private network-to-network signaling developed by the ATM Forum. Finally, Chapter 8 discusses the signaling procedures for the modification of connection characteristics (e.g., traffic parameters) after a connection is established.

Acknowledgments

We are indebted to all who contributed to this book in various ways. The PNNI signaling sections have greatly benefited from the material and review provided by Dr. Ted Tedijanto of FORE Systems. Mr. Rajiv Kapoor of AT&T, chairman of the ITU-T B-ISDN group, has reviewed several sections and made several invaluable comments on the organization of the text. Chapters 2 and 3 were reviewed by Mr. Chin Chiang, Bellcore.

Rao Cherukuri also thanks IBM management for their support and encouragement. He also extends his thanks to Carol Johnson and Jackie Putnum for their help in the preparation of the figures.

Raif O. Onvural is grateful to Kim Parolari for her contributions.

Signaling procedures reviewed in this book are due to the efforts of several people and took several years in developing. You all know who you are when we acknowledge the countless hours put out to bring us all to this point and say thank you.

1

Introduction

Signaling is the act of communication via signals in real time. A signal is a sign or means of communication agreed upon (understood) and used to convey information or a command between two or more communicating entities. In communication networks, signaling allows network users to communicate their service requirements to the network (user-to-network signaling). Similarly, it allows network devices (switches, servers, etc.) to exchange among them various types of information needed to manage the network and support user traffic (network-to-network signaling).

Not all communication networks use signaling. For example, signaling may not be used to send a packet from one workstation to another across a legacy local-area network (LAN) or a packet network with a datagram service. In these networks, in simple terms, a source end station sends its packets to a network device it is attached to. Packets are routed from one network device to another until they reach their destination(s). Each network device looks at the destination address included in the packet and uses its view of the network to decide where to send the packet next. As the network view of a network device may change over time, two packets between the same source and destination pair may follow different paths in the network.

However, not all packet-switched networks provide best-effort service only. Packet switching with virtual circuit service requires an end-to-end path to be established prior to data flow (but does not necessarily require dedicating resources for the communication to start). In this case, the connection identifiers are set up and reserved at the network access and/or in the network to uniquely identify the end-to-end connection. X.25, frame relay, and Systems Network Architecture (SNA) are examples of virtual-circuit-oriented packet-

1

switching networks. Before the user traffic can start flowing, signaling is used in these networks to request the establishment of a connection.

On the other hand, telecommunications networks always had some kind of signaling. A typical example is a telephone call. From the time that a caller hears a dial tone to the time the conversation can start, a number of signaling actions take place between the calling user and the network, within the network, and between the network and the called user. In its simplest form, signaling is used in a telephone network for the network to learn the called party's telephone number, establish a path in the network, alert the called user, and enable the conversation between the parties.

One main difference between a packet-switching network and a telecommunications network is whether or not different types of resources are reserved in the network for two (or more) end stations to communicate. In the case of a telephone call, a dedicated circuit is established between the called and the calling party before the communication can start, whereas an e-mail message is delivered using a best-effort service without any resources being dedicated to it in the network. In particular, telecommunications networks are based on circuit switching and always require a (64 Kbps) channel to be available between the communicating users (in each direction of the traffic flow) for the communication to start. The Internet uses a packet-switching network with a best-effort service to deliver packets from their sources to their destinations.

Real-time applications often require various resources to be available in the network in order for their service requirements to be met. This in turn requires signaling procedures and mechanisms to be specified for both users to access the network and between network devices to determine whether or not the required resources are available. If the required resources are available, the connection request is accepted and resources are reserved for the connection. In this context, resources may be one or more of transmission capacity, buffers, and connection identifiers. For example, the Internet Engineering Task Force (IETF) has recently completed the specification of a resource reservation protocol (RSVP) to support different applications with different service requirements in the next generation of the Internet. Asynchronous transfer mode (ATM) is a connection-oriented transfer technology and requires the use of signaling procedures between the network users and the network as well as between switches in the network.

This chapter is an overview of what signaling is, with a particular focus on ATM networks. Section 1.1 addresses the question, "Why signaling?" Formally, signaling takes place across an interface (e.g., a demarcation point between two devices). Various ATM interfaces defined by the standards organizations are reviewed in Section 1.2. The signaling structure used in ATM networks is introduced in Section 1.3. The formats of signaling messages and information elements are discussed in Sections 1.4 and 1.5, respectively. Various aspects of characterizing an ATM connection are introduced in Section 1.6. Current ATM standardization activities at different national and international standards organizations and industry forums are reviewed in Section 1.7.

1.1 Why Signaling?

For presentation purposes, let us classify network applications into four categories: (1) both delay- and loss-sensitive (e.g., interactive video), (2) delay-sensitive but tolerable to moderate losses (e.g., voice), (3) loss-sensitive but tolerable to moderate delays (e.g., interactive data), and (4) tolerant to both delay and loss (e.g., file transfer).

Consider a network that supports both delay- and loss-sensitive applications. Let us assume these applications generate traffic at a constant rate of 10 cells/s on a transmission link with a transmission rate of 100 cells/s. A first in, first out (FIFO) buffer is used to store incoming cells temporarily while a cell is being transmitted. Without considering the delay and loss requirements of applications, we may conclude that 10 such applications can be supported simultaneously at this link. If, on the other hand, there are 11 applications, each generating traffic at the constant rate of 10 cells/s, the total incoming traffic rate is greater than the transmission rate. This in turn results in buffer overflow (i.e., cell losses), independent of the buffer size.

Even with only 10 sources, it may not be possible to meet the delay and cell-loss requirements of these applications depending on their quality of service (QoS) requirements and the amount of resources available in the system. For example, if the maximum delay that can be tolerated by these applications is equal to 0.05 seconds, it is not possible to multiplex more than five sources to this link simultaneously. In particular, if there are six flows multiplexed onto this link, it is possible that six cells, one from each source, may arrive simultaneously. In this case, the delay a cell may observe varies from 0.01 to 0.06 seconds. That is, one cell incurs a delay of more than 0.05 seconds. The probability of this happening is quite low (10^{-6}), but nevertheless, it is not equal to zero.

With a buffer size of 10 cells, there will be no buffer overflows if there are 10 sources and one cell arrives every 0.1 seconds (i.e., at a constant rate of 10 cells/s) from each source. If, on the other hand, the buffer size is less than 10 cells, the probability of buffer overflow is no longer zero as all 10 sources may become active simultaneously. For example, if the buffer size is nine cells, including the cell being transmitted, the probability of a cell loss is 10^{-10}, not equal to zero.

The reality is much more complicated than what these examples present. ATM networks are designed to support a variety of applications with different service requirements and traffic requirements. However, these examples illustrate that when a network supports applications with delay and loss requirements, it is no longer possible to provide uncontrolled access to the network (i.e., best-effort service). A mechanism is needed for the end user to specify to the network its application's traffic characteristics and service requirements. Using this information, the network can (based on its internal processing and current resource utilization) decide whether or not it is possible to provide the requested service. In addition to the traffic-related information that needs to be exchanged between the end users and the network (or between two networks), end users may need to exchange some application layer information among themselves. In ATM, such information may include the ATM adaptation layer (AAL) type, the layer 3 protocol, and so forth used by the application.

In summary, supporting services with delay and loss requirements in a packet-switched network require various information to be exchanged between the end user and the network and between two networks. This information exchange takes place using signaling. More formally, signaling in a communication network is the collection of procedures used to dynamically establish, maintain, and terminate connections. For each function performed, the corresponding signaling procedures define the sequence and the formats of messages to be exchanged between the communicating entities. Each signaling message is either a request for a specific function or it is a response to a specific request. In either case, a signaling message is composed of a number of information elements. Each information element (IE) identifies a specific aspect of the function requested by the message (or the response to a request).

The set of functions performed, the messages, and the information elements in a message exchanged between communicating entities are specific to the network interface across which the exchange takes place. At the interface between the user and the network, signaling messages are used to characterize the user's connection request and its service requirements to the network and for the network to inform the user whether or not the connection request is accepted (that is, whether the system has the resources to support the connection). Similarly, between two networks or two network devices, signaling is used for the two devices to exchange connection-related information and to make connection admission decisions.

Before we proceed further, various interfaces across which signaling exchange takes place in ATM will be reviewed next.

1.2 ATM Interfaces

ATM standards provide the basic framework for the interoperability among ATM systems built by different vendors. The term "system" in this context is used generically to refer to a collection of components that comprises an end-to-end ATM solution and includes ATM end stations, ATM switches, interworking units supporting non-ATM protocols/services, and so forth.

Standardization provides users, manufacturers, and service providers the freedom of choice. Manufacturers benefit from standards, as standards give them more opportunity to compete for the business of all the potential purchasers of telecommunications equipment. Service providers or organizations that build private networks take advantage of selecting their networking equipment from multiple vendors in a cost-effective manner. Users benefit the most as standard services give them a freedom in both selecting service providers and purchasing equipment that will fit best to their networking requirements in the most cost-effective way.

ATM standards have been developed by the International Telecommunications Union-Telecommunications sector (ITU-T) and by national standards bodies such as American National Standards Institute (ANSI) and European Telecommunications Standards Institute (ETSI). ITU-T is the international authority for ATM standards. ATM Forum is not a standards organization, but it is a powerful consortium of more than 800 companies and organizations that produce interoperability agreements. ATM standards are defined based on different types of interfaces that address the connectivity and interoperability issues between the different components of ATM networks: an ATM end station and an ATM switch, two switches, service interfaces, and so forth.

A subset of interfaces that have been defined by ITU-T and ATM Forum are illustrated in Figure 1.1. ATM Forum produced its first implementation agreement in June 1992. This document, entitled User-to-Network Interface (UNI) specification version 2.0, included definitions of various physical layer interfaces and functions to support permanent ATM connections (i.e., no signaling). Since then, ATM Forum finalized UNI 3.0, UNI 3.1, and UNI 4.0 signaling specifications to support switched connections on demand via UNI signaling. User-to-network interface defined by ITU-T is referred to as public UNI. The interface between two networks (equivalently, two switches) is referred to as network-to-network interface (NNI). ITU-T's focus is on public networks, whereas ATM Forum has defined private NNI as the interface between two private switching systems. Various specifications addressing different interoperability requirements produced by ITU-T and by ATM Forum are discussed in Section 1.7.

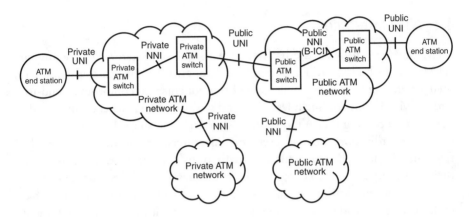

Figure 1.1 The ATM Forum interfaces.

UNI is the demarcation point between an ATM end station and an ATM network. An ATM end station in this context refers to any device that transmits ATM cells to the network. Accordingly, an ATM end station may be an inter-working unit that encapsulates data into ATM cells, an ATM switch, or an ATM workstation. Depending on whether the attached ATM network is private or public, the interface is referred to as private UNI or public UNI, respectively. If two switches are connected to each other across a UNI, and if at least one of the two switches belongs to a public network, the corresponding demarcation point is always a public UNI. The UNI specification includes the definitions of various physical interfaces, the ATM layer, the management interface, and the control plane that includes signaling procedures used across the UNI.

NNI is used in different contexts. It may be the demarcation point between either two private networks or two public networks, respectively referred to as private NNI and public NNI. An NNI may also be used between two switches. In this case, NNI is a switch-to-switch interface in private networks and a network node interface in public networks. Similar to UNI specification, NNI standards include the definitions of various physical interfaces, the ATM layer, the management interface, and signaling.

In addition to these two fundamental interfaces, ATM Forum has defined various other interfaces designed specifically to address a particular requirement. For example, LAN Emulation UNI (LUNI) is built on top of UNI signaling to allow legacy applications to run over ATM with no changes to the application. In particular, current networking applications have been developed to run over connectionless, shared media LANs such as Ethernet and Token Ring by taking advantage of the broadcast nature of the shared media. LUNI hides the connection-oriented nature of ATM and emulates a connectionless access to a shared medium over switched ATM network.

Similar to public NNI, ATM Forum's broadband intercarrier interface (B-ICI) uses public NNI signaling, broadband integrated services digital network user part (B-ISUP). The B-ICI specification includes the specification of various physical layer interfaces, ATM layer management, and higher layer functions required at the B-ICI for the interworking between ATM and various services that include switched multimegabit data service (SMDS), frame relay, circuit emulation, and cell relay. The specification also has the general principles for network provisioning, management, and accounting.

ATM data exchange interface (DXI) was developed to allow current routers to interwork with ATM networks without requiring special hardware. The DXI specification includes the definitions of the data-link protocol and physical layers that handle the data transfer between a data terminal equipment (DTE, a router) and a data communication equipment (DCE, an ATM data service unit [DSU]) as well as local management interface and management information base. DXI framework uses UNI signaling to manage switched ATM connections.

The scope of this book is limited to signaling across public UNI, private UNI, and private NNI (PNNI).

1.3 Signaling Structure in ATM Networks

ATM is a connection-oriented protocol requiring a connection to be established before user traffic can start flowing between the communicating entities. An ATM connection identifier is a 24-bit field at the UNI and a 28-bit field at the NNI included in the ATM cell header. It consists of a virtual path identifier (VPI) and a virtual circuit identifier (VCI).

On-demand connections are established in ATM using signaling. As ATM requires connections to be established prior to data flowing in the network, it is necessary to establish signaling channels before signaling messages can be exchanged between two peer entities. A signaling channel is like any other ATM layer connection except that it uses a predefined connection identifier, and they are used solely for exchanging signaling messages. There are two types of signaling channels used in ATM, associated signaling and nonassociated signaling, as illustrated in Figure 1.2.

In nonassociated signaling, the signaling entity controls all the virtual circuits (VCs) in all virtual path connections (VPCs) at its interface (except those VCs within the VPCs that are configured for use with associated signaling). Both UNI and PNNI signaling uses nonassociated signaling. The signaling virtual channel at these interfaces is identified with VPCI = 0 and VCI = 5. This is a dedicated, point-to-point ATM layer connection.

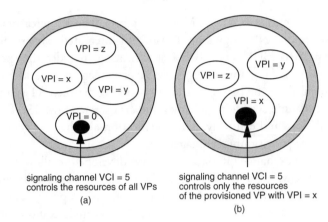

signaling channel VCI = 5
controls the resources of all VPs

(a)

signaling channel VCI = 5
controls only the resources
of the provisioned VP with VPI = x

(b)

Figure 1.2 Signaling channels across an interface: (a) associated and (b) nonassociated
signaling.

In associated signaling, the signaling entity exclusively selects and controls
all the VCs in a VPC by using a particular VC in the VPC to exchange signaling
messages. Associated signaling procedures may be used only when two sides of
an interface are connected by a VPC used as a logical link (for example, in
PNNI). A signaling channel within a VPC is identified with the identifier of the
virtual path and VCI = 5.

Both UNI and PNNI signaling refers to virtual path connection identifier
(VPCI) instead of VPI. The concept of VPCI is introduced to address the case
in which the connection between an end station and the network (or between
two networks) is through virtual path cross connects (i.e., permanent connections). When this is the case, it is necessary to provide a mapping between the
VPIs used by the end station and by the network. If there are no cross connects
at the interface, the VPCI and VPI values are the same (i.e., VPI = VPCI) and
no further mapping is required.

When cross connects are used, a virtual path between an end station and
the network (in general, between two ATM devices) may be identified differ-
ently by using different VPI values at the end station and network. In particular,
with the cross connects, an incoming VPI is mapped to a different outgoing
VPI value. Without this mapping, there would be a misassociation of a given
signaling link to a virtual path (VP).

With the cross connects at the access, a VPCI specifies a concatenation of
virtual path links extending between a device where the VCI values are assigned
and the device where those values are translated. In Figure 1.3, the cross connect
joins a VP between an end station and an ATM network with VPI = 0 on the
user side and VPI = 30 on the network side. If the network allocates VPI = 30
for a user plane connection and includes VPI = 30 in the connection identifier,

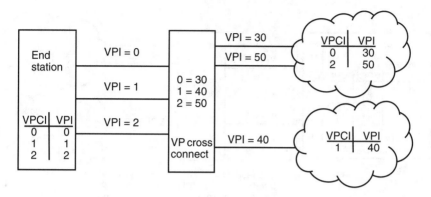

Figure 1.3 Single interface controlled by SVC.

the user will not recognize the VPI. In order to avoid this problem, the VPCI concept was introduced to provide a logical identifier for the virtual path. For example, VPCI = 0 in Figure 1.3 identifies the virtual path with VPI = 0 at the user side and VPI = 30 at the network side. The user and the network provide a mapping between VPCI and VPI values.

If the signaling virtual channel (SVC) only controls a single interface at the user side, the VPI and the VPCI have the same numerical value at the user side, as illustrated in Figure 1.3. If there is more than one signaling channel used at a physical interface, VPIs that are supported by the signaling channel have to be configured. A VPCI has significance only with respect to one particular signaling virtual channel. It is also possible to control multiple interfaces by a single signaling virtual channel. If the signaling channel controls multiple interfaces at the user side, the VPCI corresponds to both the interface and a VPI on the interface. This case is illustrated in Figure 1.4.

1.3.1 The Signaling AAL

The ATM adaptation layer (AAL) enhances the services provided by the ATM layer to support functions required by the next higher layer. Signaling in ATM networks is a layer 3 protocol and uses the services of the signaling AAL (SAAL) that comprises the AAL functions necessary to exchange signaling messages among peer entities, that is, segmentation and reassembly and reliable transfer of the signaling messages.

As illustrated in Figure 1.5, SAAL consists of AAL 5 common part convergence (CPCS) and service-specific convergence sublayers (SSCS). SAAL SSCS consists of two sublayers: service-specific coordination function (SSCF) and service-specific connection-oriented protocol (SSCOP). SAAL SSCF maps the

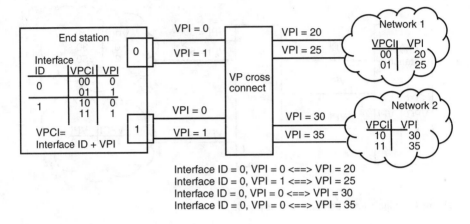

Interface ID = 0, VPI = 0 <==> VPI = 20
Interface ID = 0, VPI = 1 <==> VPI = 25
Interface ID = 0, VPI = 0 <==> VPI = 30
Interface ID = 0, VPI = 0 <==> VPI = 35

Figure 1.4 Multiple interfaces controlled by the SVC.

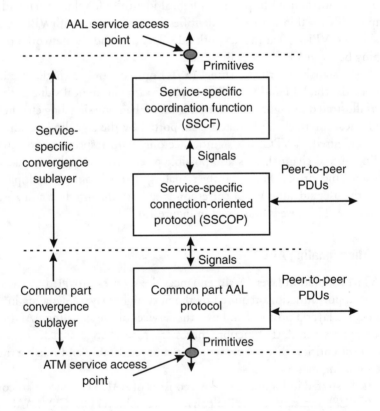

Figure 1.5 AAL structure.

particular requirements of the signaling layer to the requirements of the ATM layer, whereas SSCOP provides the mechanisms for the establishment, release, and monitoring of signaling information exchange between peer signaling entities by providing reliable transfer of signaling messages.

1.3.1.1 SSCOP

SSCOP receives variable length service data units (SDUs) from the signaling layer, then forms the protocol data units (PDUs) and transfers them to the peer SSCOP. At the receiving end, SSCOP delivers the received SDU to the signaling layer. SSCOP uses the services of CPCS. CPCS provides an unassured information transfer and a mechanism for detecting (but not correcting) corruption in SSCOP PDUs. The set of functions provided by the SSCOP are summarized as follows:

- *Sequence integrity*—preserves the order of SSCOP PDUs submitted by the signaling layer;
- *Error correction by retransmission*—receiving SSCOP detects missing PDUs, and sending SSCOP corrects sequence errors through retransmission;
- *Flow control*—allows receiver to control the rate at which peer SSCOP transmitter entity may send information;
- *Keep alive*—assures that two peer SSCOP entities participating in a connection remain in a link connection established state even in the absence of data transfer for long periods of time;
- *Local data retrieval*—allows the local SSCOP user to retrieve in sequence the SDUs that have not yet been transmitted by the SSCOP entity or those that are not yet acknowledged by the peer SSCOP entity;
- *Connection control*—permits the establishment, release, and synchronization of an SSCOP connection;
- *Transfer of user data*— used for the conveyance of user data between the users of the SSCOP;
- *Protocol control information (PCI) error detection*—detects errors in the PCIs;
- *Status reporting*—allows the transmitter and the receiver entities to exchange status information.

In order to perform these functions, different types of PDUs listed in Table 1.1 are defined for peer-to-peer communications between two SSCOP entities.

Table 1.1
SSCOP PDUs

PDU Name	Description
BEGIN	Used to initially establish an SSCOP connection or to re-establish an existing SSCOP connection
BEGIN ACKNOWLEDGE	Used to acknowledge the acceptance of an SSCOP connection request by the peer SSCOP entity
END	Used to release an SSCOP connection between two peer entities
END ACKNOWLEDGE	Used to confirm the release of an SSCOP connection that was requested by the peer SSCOP entity
BEGIN REJECT	Used to reject the connection establishment of peer entity
RESYNCHRONIZE	Used to resynchronize the buffers and the data transfer state variables in the transmit direction of a connection
RESYNCHRONIZE ACKNOWLEDGE	Used to acknowledge the resynchronization of the local receiver in response to a resynchronized PDU
SEQUENCED DATA	Used to transfer sequentially numbered PDUs containing user information
STATUS REQUEST	Used to request status information about the peer SSCOP entity
SOLICITED STATUS RESPONSE	Used to respond to a status request PDU; contains information on reception status of sequenced data, credit information for the peer transmitter, and sequence number of status request being responded to
UNSOLICITED STATUS RESPONSE	Used to respond to detection of a new missing sequenced data and credit information for the peer transmitter

1.3.1.2 SSCF

The SSCF performs a coordination function between the service required by the signaling layer and the services provided by the SSCOP. It provides a mapping between SAAL primitives and signals between the SSCF and the SSCOP. The term signal is used instead of the primitive at the SSCF and SSCOP interface in order to reflect the fact that there is no service access point between the two layers.

Various services provided by the SSCF sublayer include independence from the underlying layers, transparent relay of information, establishment of connections for assured data transfer mode, and assured data transfer mode. These capabilities are provided through a mapping between a simple state machine for the user and the more complex machine employed by the SSCOP protocol. The signals used between the SSCF and SSCOP are listed in Table 1.2.

Table 1.2
Signals Between SSCOP and SSCF at the UNI

Signal Name	Description
AA-ESTABLISH	Used to establish assured, point-to-point information between peer user entities
AA-RELEASE	Used to terminate assured, point-to-point information between peer user entities
AA-DATA	Used for the assured, point-to-point transfer of SSCOP SDUs between peer user entities
AA-RESYNC	Used to resynchronize the SSCOP connection
AA-UNITDATA	Used for the unassured transfer of SSCOP SDUs between peer user entities

1.3.2 Primitives

SAAL functions are accessed by the signaling layer through a service access point (SAAL-SAP), whereas ATM-SAP provides a bidirectional flow of information and allows ATM functions to be accessed by the AAL, as illustrated in Figure 1.5. There is always a one-to-one correspondence between a connection endpoint within the SAAL-SAP and a connection endpoint within the ATM-SAP.

In general, there are four types of primitives used to exchange data between two layers. The *request* primitive type is used when a higher layer is requesting a service from the next lower layer. The *indication* primitive type is used by a layer that provides a service to notify the next higher layer of any specific service-related activity. The indication primitive may be the result of an activity of the lower layer related to the primitive type request at the peer entity. The *response* primitive type is used by a layer to acknowledge receipt, from a lower layer, of the primitive type indication. The *confirm* primitive type is used by the layer providing the requested service to confirm that the activity has been completed.

These four primitive types, listed in Table 1.3, are used by the signaling layer to request service from the SAAL and for the SAAL to respond to these requests. The same set of primitives are also used between the SAAL and the ATM layer.

Table 1.3
UNI Signaling/SAAL Primitives

Primitive Type	Description
Request	Used when signaling layer is requesting a service from the signaling AAL layer
Indication	Used by signaling AAL layer to notify signaling layer of any service-related activity
Response	Used by signaling layer to acknowledge receipt from signaling AAL layer of the primitive type indication
Confirm	Used by the signaling AAL layer to confirm that the activity has been completed

Figure 1.6 illustrates the use of these primitive types. The signaling entity on the left requests a (specific) service from its SAAL using the request primitive type. The SAAL entity on the right uses the indication primitive type to inform its own signaling entity of a request from its peer entity. This signaling entity uses the response primitive type to indicate to its SAAL that the requested service from the SAAL is a response to a request from its peer entity. Upon receiving the corresponding message from its peer, the SAAL of the signaling entity that originated the request uses the confirm primitive to indicate that the reply to its request has arrived from the peer entity.

These primitive types are used by the signaling layer and the SAAL to indicate a specific service type together with the four function primitives listed in Table 1.4: *establish, release, data, unit-data*. The primitive itself, together with the primitive type, completely defines the action at the interface.

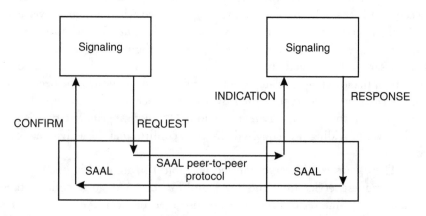

Figure 1.6 Exchange of SAAL primitives.

Table 1.4
SAAL-SAP Primitives

Primitive Name	Type	Description
AAL-ESTABLISH	Request Indication Response Confirm	Used for establishing assured information transfer between AAL entities at the UNI
AAL-RELEASE	Request Indication	Used for terminating assured information transfer between AAL entities at the UNI
AAL-DATA	Request Indication	Used in conjunction with assured data transfer at UNI
AAL-UNIT-DATA	Request Indication	Used in conjunction with unassured data transfer at UNI

Given this framework, an example of the sequence of events that take place while sending a signaling message across the UNI is illustrated in Figures 1.7 through 1.9.

Before signaling procedures are invoked, an assured-mode signaling AAL connection must be established between the user and the network.. The sequence of primitives exchanged across a UNI is illustrated in Figure 1.7.

Establishment of signaling AAL connections is initiated by transferring an AAL-ESTABLISH-request primitive to the signaling AAL. Signaling procedures may begin after receiving an AAL-ESTABLISH-confirm or AAL-ESTABLISH-indication primitive from the SAAL. The AAL-ESTABLISH-indication primitive is received in the case of SAAL establishment request by the peer en-

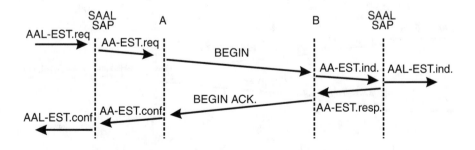

Figure 1.7 SSCOP connection establishment.

tity, and AAL-ESTABLISH-confirm in response to a local request to establish an SAAL connection.

All layer 3 messages are sent to the signaling AAL using an AAL-DATA-request primitive, as illustrated in Figure 1.8.

For example, when a signaling entity requests an ATM layer connection setup, it sends a specific signaling message to its peer at the other side of the interface. This message is carried between the two corresponding SAAL entities using an AAL-DATA protocol data unit with SAAL primitives, as illustrated in Figure 1.8. Similar to the other cases, the signaling layer informs the SAAL that it has a message to send to the other end by using the AAL-DATA-request primitive. This data is delivered at the other end from the SAAL to the signaling layer across the SAP using the AAL-DATA-indication primitive.

Figure 1.9 illustrates the sequence of primitive exchanges to release an SSCOP connection.

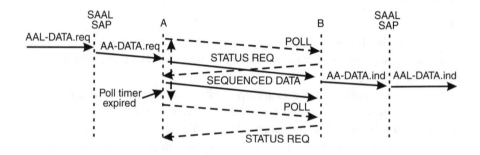

Figure 1.8 Signaling message transfer.

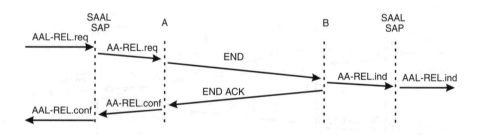

Figure 1.9 SSCOP connection release.

1.4 Signaling Message Format

As discussed previously, signaling across a UNI or PNNI takes place via exchanging signaling messages between the two peer signaling entities at the two ends of the corresponding interface. A signaling message format, illustrated in Figure 1.10, consists of the following parts: protocol discriminator, call reference, message type, message length, and variable length information elements. All of these fields are always present in a signaling message, with the only exemption being the variable length information element field, which is specific to each message type.

The same message format is used by different protocols such as ITU-T UNI signaling (i.e., Q.2931, etc.) and ATM Forum UNI signaling and PNNI signaling. The protocol discriminator is used to distinguish messages of one particular protocol from the others by assigning a unique identifier to a protocol.

The call reference field is used to identify the call at the corresponding interface to which the particular message applies. For example, each call at the UNI is assigned a unique call reference value local to each side of the interface. Similarly, each call at the PNNI is assigned a unique call reference value at the PNNI interface. The call reference is the second part of every message.

The call reference field is 4 bytes long and coded as shown in Figure 1.10. The first byte (second octet of the message) consists of the length of the call reference value field in bits 4 through 1 (0011 indicating the next three octets) whereas the higher order four bits are set to 0. The second octet of the call reference field includes one bit flag (third octet of the message, bit 8). The remaining 23 bits are the call reference value, locally unique to each call, at each interface. The only two exemptions are (1) the call reference value of 0 (all 23 bits set to 0) reserved for the global call reference and (2) the call reference value of all 1's (all 23 bits set to 1) reserved for the dummy call reference. The global call reference

Field								Octet
8	7	6	5	4	3	2	1	
Protocol discriminator								1
0	0	0	0	Length of call reference value				2
				0	0	1	1	
Flag								3
Call reference value								4–5
Message type								6–7
Message length								8–9
Variable length information elements								10–etc.

Figure 1.10 Signaling message format.

value is used to reference all calls established at an interface that are associated with a particular signaling virtual channel. The dummy call reference may be used for various supplementary services (refer to Chapter 5 for the details).

The call reference value of a call is assigned by the side (user or network) that originates the call at the interface. It is unique only at the originating side (assigned at the time the call originates) and remains fixed for the lifetime of a call. After a call ends, the associated call reference value is released and becomes available for another call. It is possible that two identical call reference values on the same signaling virtual channel may be used when each value pertains to a call originated at the opposite ends of the interface. The call reference flag value is used to address this potential conflict by identifying which end of the signaling virtual channel originated the call reference. The side that originated the call always sets the call reference flag to 0 in its messages, whereas the destination side sets it to a 1 when it replies to messages sent by the originating side.

The message type field of a signaling message is used to identify the function of the message. It is the third part of every signaling message. Figure 1.11 illustrates the format of the message type field.

Each signaling message of a protocol (i.e., UNI signaling or PNNI signaling) is assigned a unique message type value (octet 6 of the message). Octet 7 of the message type field includes the message action indicator, which allows the sender of a message to indicate explicitly the way the peer entity shall handle unrecognized messages, and a flag field. All the other bit values are set to 0 except bit 8, which is set to 1. The message action indicator has a local significance only. In general, it is a network option to which value the action indicator is set for messages sent from the network to the user. The coding of the message action indicator and the flag are defined as follows. If the flag is set to 0, the message instruction field is not significant and regular error-handling procedures applies. The flag is set to 1 to request that explicit instructions that supersede the regular error-handling procedures need to be followed. The explicit instructions may be one of the following: (1) clear call, (2) discard and ignore, (3) discard and report status, or (4) reserved.

Field								Octet
8	7	6	5	4	3	2	1	Octet
Message type								6
1 ext	0	0 Spare	Flag	Reserved	0 Spare	Message action indicator		7

Figure 1.11 Format of the message type field.

The message length field (octets 8 and 9 of the message) is used to identify the length of the message contents. It is the binary coding of the number of octets of the message, excluding the octets used for the protocol discriminator, call reference, message type, and the message length indication itself. If the message contains no further octets, the message length is coded as 0.

Different messages have different significance levels. A message is said to have a local significance if it is relevant at the local interface only. A globally significant message, on the other hand, has relevance inside the network and at interfaces that the call originates and terminates. A message with a significance at the access interface is significant only at the originating and terminating access interface, but not in the network. Finally, a message is referred to having a dual significance if it is relevant in either the originating or terminating access and in the network.

1.5 Information Elements

A message consists of a number of IEs, as shown in Figure 1.12. The specification of a message includes the list of IEs that may be included in that message. The inclusion of an IE in a message may be mandatory or optional. A particular IE may be included in more than one signaling message.

The *IE identifier* field (octet 1 of every information element) uniquely identifies the IE. The second octet of an IE consists of the coding standard and the *IE instruction* field. There are four coding standards: ITU-T standard, ISO/IEC standard, national standard, and network-specific standard (either public or private).

The *IE instruction* field consists of a flag, a reserved bit, and a 3-bit *IE action indicator* field.

The flag is used by the receiving signaling entity when it does not recognize the IE itself or any part of it. It allows the sending signaling entity to indicate whether the regular error-handling procedures apply or the processing

Field								
8	7	6	5	4	3	2	1	Octet
Information element identifier								1
1 ext	Coding standard	IE instruction field						2
		Flag	Reserved	IE action indicator field				
Length of information element parameter contents								3–4
Information element specific information								5–etc.

Figure 1.12 Information element format.

should follow the explicit instructions that supersede the regular error-handling procedures to handle unrecognized IEs defined as part of each signaling protocol.

The *reserved* bit of the IE instruction field may be used to indicate a pass along request. It is used by an upstream ATM device to request that an IE be forwarded along the path even though it may not be processed by some of the switches along the end-to-end path. When this bit is set to 0, it indicates that the IE should not be passed to the next switch along the path; whereas when it is set to 1, it indicates the IE should be passed along.

An *IE action indicator* is used to specify one of the following actions: (1) clear call; (2) discard information element and proceed; (3) discard information element, proceed, and report status; (4) discard message; or (5) discard message and report status. The value of the action indicator is, in general, a network option.

The third and fourth octets of an IE indicate the length of that IE. It is the binary coding of the number of octets of the contents of the IE specific information. The value specified in this field does not include the first four bytes of the IE (octets 1 through 4).

1.6   Characterization of an ATM Layer Connection

An ATM layer connection is established from a source end station to one or more destination end stations. Currently, ITU has specified the signaling procedures for two types of connections: point-to-point and point-to-multipoint.

A point-to-point connection is a unidirectional or bidirectional ATM connection between two end stations, whereas a point-to-multipoint connection is a unidirectional connection from a single source to two or more end stations, as illustrated in Figure 1.13.

In either case, the calling party originates the connection request to the called party. In the case of a point-to-point connection, the calling party defines

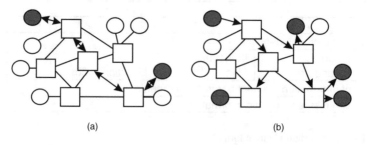

(a)   (b)

Figure 1.13   (a) Point-to-point and (b) point-to-multipoint connections in ATM.

the traffic and QoS characteristics of the connection in both directions. If the peak rate of the connection in one direction (either from the called party to the calling party or from the calling party to the called party) is zero, then the connection is a unidirectional connection (in the direction the peak rate is greater than zero); otherwise, it is a bidirectional connection (i.e., the peak rate of the connection is greater than zero in both directions).

A point-to-multipoint connection is always unidirectional. The node that originates the connection is referred to as the root, and all destination nodes are referred to as leaves. Accordingly, the traffic originates only at the root, and it is received by one or more leaves. Leaves can not use a point-to-multipoint connection to communicate with the root or with other leaves. That is, the peak rate of the connection at the interface the leaf is attached to is equal to zero in the direction from the leaf to the switch.

An extension to the basic point-to-multipoint connection is to allow the leaves to join the connection without root initiation. This feature, referred to as leaf-initiated join (LIJ), is specified in ATM Forum 4.0.

1.6.1 ATM Addresses

Each ATM end station requires a unique ATM address for ATM end stations to uniquely identify each other. Unique ATM addresses are also used by the ATM network to locate the destination end node(s) of connection requests. Finally, ATM switches themselves are assigned ATM addresses in PNNI networks to uniquely identify each switch.

Private and public networks use different ATM addressing formats. Public ATM networks use E.164 addresses (i.e., telephone numbers), whereas ATM private network addresses are based on the Open Systems Interconnection (OSI) network service access point (NSAP) format. NSAP addresses are based on the concept of hierarchical addressing domains, as illustrated in Figure 1.14. At any level of hierarchy, an initial part of the address unambiguously identifies a subdomain, and the rest is allocated by the authority associated with the subdomain to unambiguously identify either a lower level subdomain or an access point within the subdomain. Details on NSAP addresses are given in RFC1237.

Figure 1.15 illustrates the high-level format of an ATM address. A private ATM address is 20 bytes long and consists of a network addressing domain, referred to as initial domain part (IDP) and domain-specific part (DSP). IDP specifies a subdomain of the global address space and identifies the network addressing authority responsible for assigning ATM addresses in the specified subdomain. Accordingly, IDP is further subdivided into two fields: the authority and format identifier (AFI) and the initial domain identifier (IDI). AFI specifies the format of the IDI, the network addressing authority responsible for

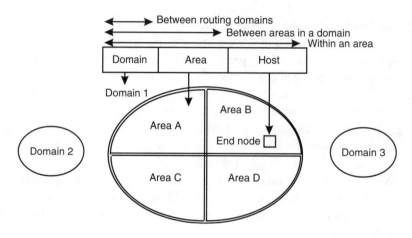

Figure 1.14 NSAP address structure.

Initial domain part (IDP)		Domain specific part (DSP)		
Authority and format identifier (AFI)	Initial domain identifier (IDI)	High order DSP	End system identifier (ESI)	Selector

Figure 1.15 Private ATM address general format.

allocating values of the IDI and the abstract syntax of the DSP. The IDI specifies the network addressing domain, from which values of the DSPs are allocated, and the network addressing authority responsible for allocating values of the DSP from that domain.

Three AFIs are defined by ATM Forum, namely the data country code (DCC), international code designator (ICD), and E.164. The specifics of the high-order DSP part is defined by the network administrator. In particular, the network part of the address can be subdivided into a number of subfields (on the bit boundary, if so desired) to create multiple levels of hierarchy in PNNI routing (cf. Chapter 7).

A private ATM address includes a 13-byte network part and a 7-byte end-station part. The end-station part is divided into two fields: a 6-byte end-station identifier and a 1-byte selector for use solely by the end station. For example, an ATM end station may support a number of terminal equipment to access an ATM network, and the selector field may be used to differentiate between different devices.

The client registration mechanism defined in UNI 3.1 allows end stations to exchange their station identifiers for the ATM address information configured at the switch port. As a result of this exchange, the end station automatically acquires the ATM network address of the switch port it is attached to without any requirement for that address to be manually provisioned into the end station. The end station then appends its own identifier and forms its full ATM address. Similarly, the end-station part of the address is registered in the network and associated with its respective network part. With this scheme, several ATM addresses with the same network-defined part (with distinct end system identifiers [ESIs]) can be registered at the network side of the UNI. It is also possible to assign more than one prefix (network portion of the address) to a UNI. Similarly, an ATM end station may require more than one network address part, if needed.

1.6.2 ATM Layer Traffic Parameters

The traffic behavior of an ATM connection is characterized to the network with two sets of parameters: (1) peak cell rate, cell delay variation tolerance and (2) sustainable cell rate, intrinsic burst size.

The peak cell rate (PCR) of a connection specifies an upper bound on the rate at which traffic can be submitted to the network. The peak rate of the traffic generated by an application may not be exactly the same as the peak rate of the connection at the network. This is due to the multiplexing of cells from different applications at the ATM end station. Let us consider two applications multiplexed at an interface with corresponding peak rates of 1/2 and 1/6 of the link transmission rate. With the transmission rate normalized to one, every sixth cell time at the link, there may be two cells waiting for transmission, as illustrated in Figure 1.16.

As a consequence, a cell may be forced to wait at times for another cell to be transmitted. This causes the PCR of the connection as observed by the net-

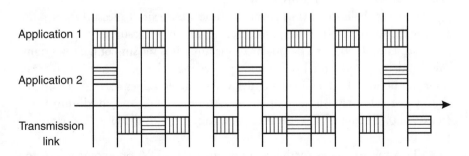

Figure 1.16 Effect of cell multiplexing to the peak cell rates of connections.

work to be different than the peak rate specified during the connection-establishment time, although the applications generate their traffic at the rate agreed upon between the network and the end station. The parameter cell delay variation tolerance (CDVT) is defined to address this problem. CDVT specifies how much the connection's instantaneous peak rate is allowed to differ from its negotiated peak rate.

The CDVT specified at the private UNI (i.e., that directly connects an end station to a private network) accounts for the cell clumping introduced by the end station. The CDVT specified at the public UNI (i.e., that connects an end system to a public network through a private ATM network via a private UNI) accounts for the cell clumping introduced by the end station and the private ATM network.

The sustainable cell rate (SCR) of a connection is an upper bound on the conforming average rate of an ATM connection, over time scales that are long relative to those for which the PCR is defined. Knowledge of the connection's average rate (more accurately, its upper bound) allows the network to allocate fewer resources in the network to the connection than would be possible when only the PCR is known, thereby allowing statistical multiplexing and better resource utilization. SCR is always defined together with the parameter intrinsic burst size (IBS). PCR, SCR, and IBS together define the maximum number of cells that can come back-to-back at the connection's PCR without violating its traffic contract.

1.6.3 ATM Layer QoS Parameters

The QoS parameters specified for ATM connections are grouped into two categories: delay parameters and dependability parameters. The delay parameters are the cell transfer delay (CTD) and cell delay variation (CDV). Dependability parameters are the cell loss ratio (CLR), cell error ratio (CER), severely errored cell block ratio, and cell misinsertion rate.

CTD is the elapsed time from the time a cell exits a measurement point (i.e., at the source UNI) to the time it enters another measurement point (e.g., the destination UNI) for a particular connection. It is the sum of the total transmission delay and the total ATM node processing delay between the two measurement points. Two end-to-end delay parameter values can be negotiated during the call establishment time: peak-to-peak CDV and maximum CTD (maxCTD). The maxCTD specified for a connection is the $(1 - \alpha)$ quantile of the CTD.

Two measurement methods are defined for CDV: one-point CDV and two-point CDV. The one-point CDV describes the variability in the pattern of cell arrival events observed at a single measurement point with reference to the negotiated peak rate 1/T (i.e., CDVT). The positive values of the one-point CDV correspond to cell clumping, whereas negative values correspond to gaps in the cell stream.

The two-point CDV describes the variability in the pattern of cell arrival events observed at the output of a connection portion with reference to the pattern of the corresponding events observed at the input to the connection portion.

The peak-to-peak CDV is the $(1 - \alpha)$ quantile of the CTD minus the fixed CTD that could be experienced by any delivered cell on a connection during the entire connection holding time. The term "peak to peak" refers to the difference between the best and worst case of CTD, where the best case is equal to the fixed delay and the worst case is equal to a value likely to be exceeded with probability no greater than α. Networks have a limited ability to control peak-to-peak CDV. Therefore, end systems cannot expect to negotiate arbitrarily small values of peak-to-peak CDV as their sole means of meeting jitter and wander tolerances.

The accumulation algorithms provide estimates of the end-to-end values of these parameters along a path. A simple accumulation based on worst case assumptions is supported in the current signaling specifications.

In the simple peak-to-peak CDV accumulation, a switch receives the accumulated peak-to-peak CDV and adds its contribution of the peak-to-peak $CDV(\alpha)$ to the accumulated peak-to-peak CDV. Similarly, in the case of simple accumulation for maxCTD, a switch receives the accumulated maxCTD and adds its contribution of the $maxCTD(\alpha)$ to the accumulated maxCTD. In signaling, maxCTD is accumulated only in the forward direction (maxCTDF). However, the CDV is accumulated in both forward and backward directions (CDV_F, CDV_B). Consequently the maxCTD for the backward direction is derivable as

$$maxCTD_B = CDV_B + maxCTD_F - CDV_F$$

This is because the fixed delay in the forward direction $(maxCTD_F - CDV_F)$ is assumed to be the same as the fixed delay in the reverse direction $(maxCTD_B - DV_B)$.

Cell loss ratio (CLR) is the only end-to-end dependability parameter currently specified in UNI and PNNI signaling. CLR is defined as the ratio of number of lost cells to the total number of transmitted cells. The CLR parameter is the value of CLR that the network agrees to offer as an objective over the lifetime of the connection.

This parameter is not explicitly accumulated. Each network element accepts or rejects the call based on a comparison between the loss rate supported by the network element and the requested CLR.

The other dependability parameters that are not signaled but affect the service quality of an ATM connection are defined in Table 1.5.

Table 1.5
Dependability Parameters

Parameter	Description
Cell error ratio (CER)	Ratio of the errored cells to the sum of successfully transferred cells and errored cells
Severely errored cell block ratio (SECBR)	Ratio of the severely errored cell blocks to the total number of transmitted cell blocks. A cell block is a sequence of N cells transmitted consecutively on a given connection. A severely errored cell block outcome occurs when more than M errored cells, lost cells, or misinserted cell outcome are observed in a received cell block.
Cell misinsertion rate (CMR)	Ratio of the number of misinserted cells to the time interval. Cell misinsertion on a particular connection is most often caused by an undetected error in the header of a cell being transmitted on a different connection.

1.6.4 ATM Transfer Capabilities

ITU-T specified a set of ATM transfer capabilities (ATCs) that are designed to support a variety of ATM layer service models with associated QoS capabilities through a set of ATM layer traffic parameters and procedures. The set of currently defined ATCs address the requirements of a large number of applications. The specifications of different transfer capabilities also provide the basic framework for the network operator to meet the service requirements of different applications with different service requirements and traffic parameters while achieving high utilization of network resources.

The currently defined ATCs are classified into four categories as listed in Table 1.6.

Table 1.6
ITU-T ATM Transfer Capabilities

Transfer Capability	Description
Deterministic bit rate (DBR)	Designed to support connections that request a continuously available, fixed amount of bandwidth as long as the connection is active. The amount of bandwidth dedicated for an ATM connection is specified by the connection's peak cell rate value.
Statistical bit rate (SBR)	Designed to provide a basic framework for the network to support applications that generate traffic at a varying rate so that if the network takes advantage of this type of information, it can achieve higher utilization of its resources.
ATM block transfer (ABT)	An ATM block, in this context, is a group of cells from an ATM connection delineated by two resource management (RM) cells: a leading RM cell preceding the first cell of the ATM block and a trailing RM cell that follows the last cell of the ATM block. The ATM layer traffic characteristics are negotiated on an ATM block basis.
Available bit rate (ABR)	Designed to support applications that have the ability to adjust their transfer rate based on bandwidth availability in the network. In this transfer capability, a mechanism is defined for the network to provide feedback to ABR sources on the availability of its resources.

ITU-T ATCs are referred to as service categories in ATM Forum TM 4.0. Some of the transfer capabilities and service categories are the same, whereas some others are different. In particular, ATM Forum defines constant bit rate (CBR), variable bit rate (VBR) services, unspecified bit rate (UBR) services, and available bit rate (ABR) services. DBR is equivalent to CBR, whereas SBR is equivalent to VBR. ATM Forum's UBR service has no equivalent ITU-T transfer capability. Although the ABR service definition has not yet been finalized by ITU-T, it is expected to be the same as ATM Forum's ABR service definition.

1.6.5 Multiconnection Call

The ITU-T ATM call model includes separation of call and connection. An ATM call can include multiple connections or even no connection. The concept of connection is similar to virtual connection. But a call between two users becomes an association that describes a set of connections between them.

By extending the call between two users to support multiple connections, multimedia applications can be supported more effectively. Since each media type of a multimedia communications application has its own communications requirements (bandwidth and QoS), it is desirable to carry each media type over a separate VC. These connections can be set up, modified, and torn down in parallel. This provides flexibility and performance improvements, as for example, tearing down two or more connections of a call individually would impose a larger signaling load to the network whereas all connections can be released by releasing one call.

The call without connection allows prenegotiation of endpoint resources without actually setting up the connections. It also allows the setting of a call with one connection and addition of connections later. This can be easily extended to multiparty calls. ITU is currently working on defining multiconnections for point-to-point calls.

1.7 ATM Standards

Standardization is necessary and provides the basic architectural and operational framework for building communication networks. The ATM standards have been developed by ITU-T with contributions from various national standardization organizations such as ANSI and ETSI. The ATM Forum, on the other hand, is a consortium of more than 800 companies established to speed up the development and deployment of ATM products through interoperability specifications. This section describes these organizations and provides a summary of various ATM standards and specifications produced by them.

1.7.1 International Telecommunications Union

ITU is a United Nations agency with activities including the regulation, standardization, and development of international telecommunications. ITU is organized into three sectors reflecting its main activities:

- ITU Telecommunication Standardization Sector (ITU-T)
- ITU Radiocommunication Sector (ITU-R)

- ITU Telecommunication Development Sector (ITU-D)

ITU-R is responsible for international radio regulation and for recommendations on technical and operational matters in radiocommunication.

ITU-D's aim is to contribute to the growth and development of telecommunications throughout the world, with particular emphasis on the requirements of developing countries.

ITU-T (formerly CCITT) is the body that studies technical, operating, and tariff questions and adopts recommendations on these areas with a view to standardizing telecommunications on a worldwide basis. Where telecommunications technical standards are concerned, ITU-T is clearly the standards body. ATM development was started by this body.

Various recommendations that have been developed by ITU-T include the following:

- General network planning, network operation, and network architecture;
- Terminals (e.g., data modems, multiplexers), systems (e.g., switching, signaling, and transmission), and networks (e.g., telephone, data, ISDN, B-ISDN);
- Services and applications (e.g., multimedia, network, and service interworking);
- Operations and maintenance [e.g., telecommunication management network (TMN)];
- Tariff principles and accounting rates.

In order to achieve its objectives, ITU-T works through world telecommunication standardization conferences, advisory groups, standardization bureaus, and standardization study groups.

Standardization work is done in study groups. Currently, there are 15 study groups exploring different aspects of telecommunications, as illustrated in Table 1.7.

Each study group is divided into a number of working parties (WPs), each of which has a specific area of work. Working parties are further divided into expert teams, ad hoc groups, and so on. For example, there are four working parties within study group 13.

- WP-1: Networking capabilities and internetworking;
- WP-2: B-ISDN (ATM) aspects;

Table 1.7
Study Groups of the Standardization Sector

Study Group	Title
2	Network and service operation
3	Tariff and accounting principles
4	TMN (telecommunication management network) and network maintenance
5	Protection against electromagnetic environment effects
6	Outside plant
7	Data networks and open system communications
8	Characteristics for telematic services
9	Television and sound transmission
10	Languages and general software aspects for telecommunications applications
11	Signaling requirements and protocols
12	End-to-end transmission performance of networks and terminals
13	General network aspects
15	Transport networks, multimedia services systems and equipment

- WP-3: Transport networks and layer 1 access (physical layer);
- WP-4: Performance aspects.

Furthermore, WP-2 us subdivided into six subworking parties (SWPs) as follows:

- SWP-1: ATM layer;
- SWP-2: AAL layer;
- SWP-3: Operations and Maintenance (OAM);
- SWP-4: Traffic management;
- SWP-5: Connectionless data service;
- SWP-6: Integrated video services (IVS).

Various recommendations that have been produced by ITU-T are summarized as follows.

1.7.1.1 ITU-T Recommendations

ITU-T work on ATM includes a large variety of features at both user-to-network and network-to-network interfaces for public networks. ITU-T recommendations address the B-ISDN architecture, including the specifications of user, control, and management planes, and the physical, ATM, and ATM adaptation layers used by these planes.

The B-ISDN architectural framework is defined in the I series of recommendations listed in Table 1.8. UNI signaling of the control plane is specified in the Q series of recommendations.

Table 1.8
ITU-T I Series of Recommendations

Recommendation	Scope
I.113	*Vocabulary of terms for broadband aspects of ISDN* Provides the definitions of terms considered essential in understanding the principles of B-ISDN
I.121	*Broadband aspects of ISDN* Presents the basic principles of B-ISDN
I.151	*B-ISDN asynchronous transfer mode functional characteristics* Presents a high-level review of the ATM layer functions
I.211	*B-ISDN service aspects* Classifies broadband services into a number of service classes and presents examples of services in each class
I.311	*B-ISDN general network aspects* Defines and presents a high-level review of ATM transport network hierarchy and the specification of each layer, including signaling
I.321	*B-ISDN protocol reference model and its applications* Defines the B-ISDN reference architecture
I.327	*B-ISDN functional architecture* Presents a review of B-ISDN functions
I.356	*Quality of service configuration and principles* Defines parameters for quantifying the ATM call transfer performance of a B-ISDN connection
I.361	*B-ISDN ATM layer specification* Addresses the ATM cell structure, header coding, and the ATM protocol structure

Table 1.8 (Continued)

Recommendation	Scope
I.362	*B-ISDN ATM adaptation layer (AAL) functional description* Presents the AAL functional organization and the basic principles of AALs
I.363	*B-ISDN ATM adaptation layer (AAL) specification* Provides the details of various types of AALs defined to support different B-ISDN service classes
I.364	*Support of broadband connectionless data service on B-ISDN* Describes the support of a connectionless data service based on AAL type 3/4 on B-ISDN
I.371	*Traffic control and congestion control in B-ISDN* Presents the objectives and mechanisms of traffic control and congestion control; defines traffic descriptors, traffic parameters, and traffic contract
I.374	*Network capabilities to support for multimedia* Identifies network capabilities to support multimedia services (involving at least two of voice, image, video, and data)
I.413	*B-ISDN user-network interface* Presents the reference configuration for the B-ISDN user-network interface and various examples of physical realizations
I.432	*B-ISDN user-network interface - physical layer specification* Defines the physical media and the transmission system, and discusses the implementation of related OAM functions at the user-network interface
I.555	*Interworking* Provides the functional requirements and configuration for interworking between frame relay and other services
I.610	*B-ISDN operation and maintenance principles and functions* Identifies the minimum set of functions required to operate and maintain physical layer and ATM layer aspects of B-ISDN user-network interface as well as the individual virtual path and virtual channel connections routed through a B-ISDN

1.7.1.2   ITU-T Recommendations on B-ISDN Signaling

Various capabilities supported by B-ISDN Capability Set 2 (CS-2) are based on a concept of two connection types:

1. Connection type 1 (point-to-point connection)—A unidirectional or bidirectional ATM connection between two end stations;
2. Connection type 2 (point-to-multipoint connection)—A unidirectional connection from a single source to one or more end stations.

CS-2 provides functions for the support of both point-to-point and point-to-multipoint connections. The current list of recommendations for CS-2 signaling capability includes the following.

Q.2100: Q.SAAL.0 B-ISDN Signaling ATM Adaptation Layer Overview Description

This recommendation describes the different components that make up the AAL functions necessary to support B-ISDN signaling.

Q.2110: Q.SAAL.1 SSCOP Specification

This recommendation specifies the peer-to-peer protocol for the transfer of information and control between any pair of SSCOP entities.

Q.2120: Meta-Signaling

This recommendation defines the meta-signaling protocol used to establish and maintain signaling connections at the user-network interface.

Q.2130: Q.SAAL.2 B-ISDN Signaling ATM Adaptation Layer—SSCF for the Support of Signaling at the UNI

This recommendation defines a mapping between the SSCOP of the signaling AAL and the Q.2931 layer at the user-network interface.

Q.2931: B-ISDN Signaling

This recommendation defines the signaling procedures and functions used at a user-network interface.

Q.2932: Generic Functional Support of B-ISDN Supplementary Services

The generic protocol functions defined in this recommendation provide a means of exchanging ROSE (remote operations service element) components on behalf of signaling application in peer entities. These signaling applications may either be for the support of supplementary services or other features such as look ahead and status request.

Q.2951: Number Identification Supplementary Services

This recommendation specifies the signaling procedures to support various supplementary services such as direct dial-in, multiple subscriber number, calling line identification presentation, calling line identification restriction, connected line identification presentation, connected line identification restriction, and subaddressing.

Q.2957: Additional Information Transfer Supplementary Services

This recommendation specifies the signaling procedures to support user-to-user signaling supplementary service. This service allows a user to send/receive a limited amount of information to/from another user over the signaling virtual channel.

Q.2959: Call Priority

This recommendation specifies the messages, information elements, and procedures needed to specify different priorities to different connections. These procedures allow a user to provide to the network the priority information for each call setup request.

Q.2961: Additional Traffic and QoS Parameter Indications

Capability set 1 (CS-1) included a limited set of traffic and QoS parameter indications. CS-2 extends this parameter set to allow users to specify their service requirements more accurately and for the network to achieve better resource utilization through statistical multiplexing. In addition to the PCR defined in CS-1, the CS-2 traffic parameters include the SCR and maximum burst size (MBS). ABR and ATM transfer capabilities are also supported in CS-2. Q.2961 specifies the protocol procedures, formats, and functions needed to support these additional traffic parameters and transfer capabilities. In particular, Q.2961.1 specifies SCR and MBS for variable bit rate, and Q.2961.2 specifies ATCs.

Q.2962: Negotiation of Traffic and QoS Parameters During Call/Connection Establishment

This recommendation specifies the signaling protocol used to negotiate the traffic and QoS characteristics of connections for point-to-point and for the first party of point-to-multipoint call/connections. The negotiation takes place by using an alternative traffic descriptor and a minimum traffic descriptor. The minimum traffic descriptor allows the specification of a range of values for the PCR only. The alternative ATM traffic descriptor allows the negotiation of all traffic parameters.

Q.2963: Renegotiation/Modification of Traffic and QoS Parameters for Already Established Connections

This recommendation specifies the signaling protocol to modify the PCR of an active (already established) point-to-point call/connection under the control of the user.

Q.2964: B-ISDN Look Ahead

The look ahead procedure allows an ATM network to check whether or not the addressed end station is compatible and whether or not the end station is free to accept the connection. This procedure may be used prior to call/connection establishment. The look ahead can be initiated only by the network.

Q.2971: Support of Point-to-Multipoint Connections

This recommendation specifies the signaling procedures for switched point-to-multipoint connections from a single end station to multiple end stations. These procedures define the capability to add or remove end stations to and from a point-to-multipoint connection.

1.7.1.3 ITU-T Recommendations on Supplementary Services

Table 1.9 lists ITU-T recommendations that define various supplementary services.

Table 1.9
ITU-T Recommendations on Supplementary Services

Recommendation	Title
I.130 (1988)	Method for the characterization of telecommunication services supported by an ISDN and network capabilities of an ISDN
E.164 (1991)	Numbering plan for the ISDN era
I.580 (1993)	General arrangements for interworking between B-ISDN and 64-Kbps-based ISDN
I.251.1 (1992)	Direct dialing-in
I.251.2 (1992)	Multiple subscriber number
I.251.4 (1992)	Calling line identification presentation
I.251.4 (1992)	Calling line identification restriction
I.251.5 (1988)	Connected line identification presentation

Table 1.9 (Continued)

Recommendation	Title
I.251.6 (1988)	Connected line identification restriction
Q.2951 (1995)	Description of the number identification supplementary service using DSS-2 basic call
I.251.8 (1992)	Subaddressing supplementary service
I.257.1 (1992)	User-to-user signaling
Q.2957 (1995)	Description for additional information transfer supplementary services using DSS-2 basic call, clause 1- user-to-user signaling

1.7.2 ATM Forum

ATM Forum is an international nonprofit organization formed in October 1991. Current membership exceeds 800 organizations worldwide—computer vendors, LAN and WAN vendors, switch vendors, local and long-distance carriers, government and research agencies, and potential ATM users.

The main mission of ATM Forum is to speed up the development and deployment of ATM products through interoperability specifications. Accordingly, ATM Forum is not a standards organization. Instead, it produces implementation agreements based on international standards, where standards are available. In other words, early deployment of ATM products requires that specifications be available much earlier than the target dates of standards bodies and the goal of ATM Forum is to fill specification gaps, select options, and set parameters produced by international standards.

ATM Forum cannot take its specifications to any standards organizations as contributions. Instead, Forum specifications have been contributed to various standards bodies by the member companies who are also members of national standards bodies. However, there is always the possibility that Forum specifications will be incompatible with international standards and become a de facto standard.

ATM Forum consists of a worldwide technical committee; three marketing committees for North America, Europe, and Asia Pacific; and the Enterprise Network Roundtable, as listed in Table 1.10.

The Technical Committee works with other standards bodies, such as ANSI and ITU-T, in selecting appropriate standards, resolving differences among standards, and recommending new standards when existing ones are

Table 1.10

The ATM Forum Organization

ATM Forum Committee	Functions
Technical Committee	Produces interoperability specifications
Market Awareness Committees	Promote ATM technology within both the industry and the end-user community:
North American European Asia Pacific	Design and promote end-user interaction Raise public awareness of the ATM technology (and ATM Forum) Publicize the efforts of ATM Forum through news releases
Enterprise Network Roundtable	A user group that provides feedback and input to the Technical Committee in their efforts to develop Forum specifications in response to well understood and analyzed multi-industry requirements

absent or inappropriate. It consists of several working groups that investigate different areas of ATM products and services.

The ATM Market Awareness Committees provide marketing and educational services designed to speed up the understanding and acceptance of ATM technology. In particular, this committee coordinates the development of educational presentation modules and technology papers, facilitates exchange of information and requirements between the Enterprise Network Roundtable and the Technical Committee, publishes *53 Bytes* (the ATM Forum newsletter), coordinates publicity of Forum activities, and coordinates demonstrations of ATM at trade shows, highlighting ATM's ability to solve various business problems.

The Enterprise Network Roundtable consists of ATM end users. This group interacts regularly with the Market Awareness Committees to ensure that ATM Forum technical specifications meet the real world end-user needs.

All interoperability specifications are produced by the subworking groups of the Technical Committee. Currently, there are 12 subworking groups: signaling, broadband intercarrier interface, physical layer, traffic management, private NNI, LAN Emulation, service aspects and applications, network management, security, wireless ATM, residential broadband, and testing.

The B-ICI group defines a carrier-to-carrier interface to provide a basic framework to facilitate end-to-end national and international carrier service. This requires the specification of various physical layer interfaces and the protocols and procedures to support the transport and multiplexing of multiple services for intercarrier delivery.

The main focus of the physical layer group is the development of specifications for ATM transmission on different types of transmission mediums, including fiber, unshielded twisted pair, shielded twisted pair, coax, and copper.

The PNNI group is defining the private switching-system-to-switching-system interface in which a switching system may consist of a single switch (switch-to-switch interface) or may be a subnetwork (network-to-network interface). The work scope includes PNNI signaling and PNNI routing frameworks.

The network management group is focused on the specification of managed objects in ATM networks and information flows between management systems based on existing standards whenever they are available.

The service aspects and applications (SAA) group is chartered to define specifications that enable new and existing applications such as audiovisual services and circuit emulation to use AAL services.

Traffic management groups works on traffic aspects of ATM networking that include specifications of application traffic parameters, conformance of user traffic, development of quality of service guidelines, and definitions of service classes.

Based on specifications produced by other groups, the testing group works on designing interoperability, conformance, and performance test suites.

The LAN Emulation group is working towards defining a LAN Emulation architecture to emulate connectionless service required to support existing LAN applications without any changes over connection-oriented ATM networks.

Finally, the signaling group works on additional features, procedures, and functions needed to address the signaling requirements of existing and emerging high-bandwidth applications. The latest specification produced by this group is the ATM Forum UNI signaling interoperability agreement (UNI 4.0)

1.7.3 American National Standards Institute

Exchange Carriers Standards Association (ECSA) is the ANSI accredited body for developing standards and technical reports related to interfaces for U.S. telecommunications networks. Its working group is called Committee T1-Telecommunications. T1 develops positions on related subjects under consideration in ITU-T. Specifically, T1 focuses on those functions and characteristics associated with interconnection and interoperability of telecommunications networks at interfaces with end users, carriers, and information and enhanced service providers. To carry out its work, ECSA has established six technical subcommittees, as listed in Table 1.11.

Table 1.11
Committee T1

Technical Subcommittee	Responsibilities
T1A1	Performance and signal processing
T1E1	Network interfaces and environmental considerations
T1M1	Internetwork operations, administration, maintenance, and provisioning
T1P1	Systems engineering, standards planning, and program management
T1S1	Service architecture and signaling
T1X1	Digital hierarchy and synchronization

Various approved ANSI standards include those listed in Table 1.12.

Table 1.12
ANSI Standards on ATM

Standard	Scope
T1.624.1993	*B-ISDN UNI: Rates and formats specification* Defines the mapping of ATM cells into DS-3 and SONET payloads
T1.627.1993	*B-ISDN ATM functionality and specification* Defines the ATM layer (following I.361), including extended interpretations of traffic management and further explanations of the protocol model
T1.629.1993	*B-ISDN ATM adaptation layer 3/4 common part functionality and specification* Defines the AAL 3/4 functionality (following I.363)
T1.630.1993	*B-ISDN adaptation layer for constant bit rate services functionality and specification* Defines AAL-1 (following I.363), including specifics of emulating the North American DS-1 circuit function, interface, and management
T1.633	*Frame relay bearer service interworking* (following I.555)
T1.634	*Frame relay service specific convergence sublayer* (following I.365)
T1.635	*B-ISDN ATM adaptation layer type 5* Defines AAL-5 (following I.363)

1.7.4 European Telecommunication Standards Institute

ETSI produces standards for the European telecommunications market. The activities of ETSI are aimed at building upon ITU-T standards. The 11 ETSI technical committees have the following areas of focus:

- Network aspects;
- Business telecommunications;
- Signaling protocols and switching;
- Transmission and multiplexing;
- Terminal equipment;
- Equipment engineering;
- Radio equipment and systems;
- Special mobile group;
- Paging systems;
- Satellite Earth stations;
- Advanced testing methods.

2

UNI Point-to-Point Signaling

UNI is the demarcation point between an end station and the network. In this context, an end station can be a desktop, a switch, a gateway, and so forth, that is capable of running ATM protocols. ITU-T Recommendation Q.2931 specifies the procedures used to establish, maintain, and clear switched point-to-point connections at the UNI. ATM Forum UNI signaling specifications versions 3.1 (UNI 3.1) and 4.0 (UNI 4.0) are based on Q.2931, with various extensions to support features that may be used in private ATM networks.

ATM is a connection-oriented transfer mode and requires ATM layer connections to be established prior to traffic flowing between the communicating end stations. Each ATM layer connection is bidirectional, and the connection characteristics can be different in each direction. For example, a connection between a TV and a video server may require only a few tens of kilobits per second of bandwidth from the TV to the server, whereas a few megabits per second of bandwidth may be needed from the server to the TV. Different applications have different service requirements. Q.2931 also allows different QoS classes to be requested in each direction. A connection's QoS requirement is specified using a QoS class or individual QoS parameters appropriate for network application. In particular, UNI 3.1 supports only the QoS class, whereas the QoS parameter support is defined in UNI 4.0.

Signaling procedures are used to manage on-demand, switched virtual channel connections in real time. Once established, a switched connection can remain active for an arbitrary amount of time. The end station that originates the connection setup request is referred to as the calling party, and the destination end station is referred to as the called party. The direction from the calling party to the called party is referred to as the forward direction, and the direction from the called party to the calling party is referred to as the backward direction.

In general, a call may consist of one or more connections. For example, a video call may require one connection for voice, another for video, and one other for data. A voice call, on the other hand, may require only one connection. Q.2931 procedures are designed to support a single connection per call. Mechanisms to support multiple connections per call are currently under development in ITU SG 11 and will be contained in the Q.298x series of recommendations.

Q.2931 uses out-of-band signaling. That is, a specific ATM layer connection with a specific virtual path identifier/virtual circuit identifier (VPI/VCI) value is used to carry signaling messages. This signaling channel is different from the connections used for exchanging data between the end parties. In particular, VCI = 5 is reserved in every VPI for signaling and cannot be used to carry user traffic.

2.1 Overview of Call/Connection Control

A Q.2931 signaling entity requests and accepts services from the signaling ATM adaptation layer (SAAL) via service primitives. The primitives represent, in an abstract way, the logical exchange of information and the control between a Q.2931 entity and the SAAL. As discussed in Chapter 1, four types of primitives are used across this interface: *request, indication, response,* and *confirm.* These primitives are used by the signaling layer and the SAAL to indicate one of the following specific service types: *establish, release, data,* and *unit data.* A primitive together with a primitive type completely defines the action at the interface. The primitives exchanged between a Q.2931 signaling entity and the SAAL are summarized in Table 2.1.

Throughout this chapter, incoming and outgoing are used to describe the call as viewed by the user side of the interface, as illustrated in Figure 2.1.

Figure 2.1 Incoming and outgoing calls.

Table 2.1
Primitives between Q.2931 and SAAL

Primitive	Description
AAL-ESTABLISH-request	Issued by a Q.2931 entity to the SAAL to request establishment of an assured SAAL connection between a user and the network
AAL-ESTABLISH-indication	Issued by the SAAL to a Q.2931 entity to inform it that it has established an assured SAAL connection between a user and the network, due to a request issued by the peer Q.2931 entity
AAL-ESTABLISH-confirm	Issued by the SAAL to a Q.2931 entity to inform it that it has established an assured SAAL connection between a user and the network. This is a response to a signaling AAL establishment request by the Q.2931 entity.
AAL-RELEASE-request	Issued by a Q.2931 entity to the SAAL to request release of an assured SAAL connection
AAL-RELEASE-indication	Issued by the SAAL to a Q.2931 entity to inform it that it has released an assured SAAL connection, due to a release request issued by the peer Q.2931 entity or due to an error
AAL-RELEASE-confirm	Issued by the SAAL to a Q.2931 entity to inform it that it has released an assured SAAL connection between a user and the network. This is a response to signaling AAL release request by the Q.2931 entity.
AAL-DATA-request	Issued by a Q.2931 entity to the SAAL to request it to send a message across an established assured SAAL connection
AAL-DATA-indication	Issued by the SAAL to a Q.2931 entity to deliver it a message that was sent by a peer Q.2931 entity using an AAL-DATA-request primitive
AAL-UNIT_DATA-request	Issued by a Q.2931 entity to the SAAL to request it to send a message to one or more peer Q.2931 entities. It is sent using unacknowledged, unassured data transfer.
AAL-UNIT_DATA-indication	Issued by the SAAL to a Q.2931 entity to deliver it a message that was sent by a peer Q.2931 entity using an AAL-UNIT_DATA-request primitive

The primitives exchanged between the call control and Q.2931 signaling entities are summarized in Tables 2.2 and 2.3.

Table 2.2
Primitives From Call Control to Q.2931 Signaling Entity

Primitive	Description
Setup request	Issued by call control to Q.2931 entity to request establishment of a call/connection
Proceeding request	Issued by call control to Q.2931 entity to initiate sending a CALL PROCEEDING message
Alerting request	Issued by call control to Q.2931 entity to initiate sending an ALERTING message
Setup response	Issued by call control to Q.2931 entity to initiate sending a CONNECT message accepting the call/connection
Setup complete request	Issued by call control to Q.2931 entity to initiate sending a CONNECT ACK message
Release request	Issued by call control to Q.2931 entity to send a RELEASE message and initiate call clearing
Release response	Issued by call control to Q.2931 entity indicating that call clearing is complete and send a RELEASE COMPLETE message

Table 2.3
Primitives From Q.2931 Signaling Entity to Call Control

Primitive	Description
Setup indication	Sent by Q.2931 entity to call control to indicate receipt of a SETUP message for call/connection establishment
Proceeding indication	Sent by Q.2931 entity to call control indicating the receipt of a CALL PROCEEDING message
Alerting indication	Sent by Q.2931 entity to call control indicating the receipt of an ALERTING message
Setup confirm	Sent by Q.2931 entity to call control indicating that the connection establishment is completed and the CONNECT message is received
Release indication	Sent by Q.2931 entity to call control indicating the receipt of a RELEASE message and to initiate call clearing
Release confirm	Sent by Q.2931 entity to call control indicating that the call clearing is complete and a RELEASE COMPLETE message is received

The UNI signaling protocol is described using a call state machine. Call/connection states used in Q.2931 are listed in Tables 2.4 and 2.5. The state machines for various signaling functions are described throughout the chapter.

Table 2.4
Call/Connection States at the User Side of the Interface

State	Description
Null (U0)	No call exists
Call initiated (U1)	Exists for an outgoing call when the user requests call establishment from the network
Outgoing call proceeding (U3)	Exists for an outgoing call when the user has received acknowledgment that the network has received all call information necessary to effect call establishment
Call delivered (U4)	Exists for an outgoing call when the calling user has received an indication that remote user alerting has been initiated
Call present (U6)	Exists for an incoming call when the user has received a call establishment request but has not yet responded
Call received (U7)	Exists for an incoming call when the user has indicated alerting but has not yet answered
Connect request (U8)	Exists for an incoming call when the user has answered the call and is waiting to be awarded the call
Incoming call proceeding (U9)	Exists for an incoming call when the user has sent acknowledgment that the user has received all call information necessary to effect call establishment
Active (U10)	Exists for an incoming call when the user has received an acknowledgment from the network that the user has been awarded the call. This state exists for an outgoing call when the user has received an indication that the remote user has answered the call.
Release request (U11)	Exists when the user has requested the network to clear the end-to-end connection (if any) and is waiting for a response
Release indication (U12)	Exists when the user has received an invitation to disconnect because the network has disconnected the end-to-end connection (if any)

Table 2.5
Call/Connection States at the Network Side of the Interface

State	Description
Null (N0)	No call exists
Call initiated (N1)	Exists for an outgoing call when the network has received a call establishment request but has not yet responded
Outgoing call proceeding (N3)	Exists for an outgoing call when the network has sent acknowledgment that the network has received all call information necessary to effect call establishment
Call delivered (N4)	Exists for an outgoing call when the network has indicated that remote user alerting has been initiated
Call present (N6)	Exists for an incoming call when the network has sent a call establishment request but not yet received a satisfactory response
Call received (N7)	Exists for an incoming call when the network has received an indication that the user is alerting but has not yet received an answer
Connect request (N8)	Exists for an incoming call when the network has received an answer but the network has not yet awarded the call
Incoming call proceeding (N9)	Exists for an incoming call when the network has received acknowledgment that the user has received all call information necessary to effect call establishment
Active (N10)	Exists for an incoming call when the network has awarded the call to the called user. This state exists for an outgoing call when the network has indicated that the remote user has answered the call.
Release request (N11)	Exists when the network has received a request from the user to clear the end-to-end connection (if any)
Release indication (N12)	Exists when the network has disconnected the end-to-end connection (if any) and has sent an invitation to disconnect the user-network connection

Tables 2.6 and 2.7 define the states that Q.2931 may adopt using the global call reference. There is only one global call reference value per signaling virtual channel.

Table 2.6
B-ISDN Call/Connection States for Global Call Reference at the User Side of the User-Network Interface

State	Description
Null (Rest 0)	No transaction exists
Restart request (Rest 1)	Exists for a restart transaction when the user has sent a restart request but has not yet received an acknowledgment response from the network
Restart (Rest 2)	Exists when a request for a restart has been received from the network and responses have not yet been received from all locally active call references

Table 2.7
B-ISDN Call/Connection States for Global Call Reference at the Network Side of the User-Network Interface

State	Description
Null (Rest 0)	No transaction exists
Restart request (Rest 1)	Exists for a restart transaction when the network has sent a restart request but has not yet received an acknowledgment response from the user
Restart (Rest 2)	Exists when a request for a restart has been received from the user and a response has not yet been received from all locally active call references

2.2 Information Elements

The IEs used by various Q.2931 call setup messages are reviewed in this section, and the details of each Q.2931 message are described in subsequent sections.

2.2.1 ATM Adaptation Layer Parameter Information Element

AAL parameter IE is used to exchange/negotiate AAL parameter values between the end stations. It has an end-to-end significance only, and its contents are normally transparent to the network except in the case of internetworking. The format of the AAL parameter IE is illustrated in Figure 2.2.

Field	Octet
AAL type	5
Further contents depend on the AAL type	6 etc.

Figure 2.2 ATM adaptation layer parameter information element format.

Currently, four types of AALs are supported in Q.2931: AAL 1, AAL 3/4, AAL 5, and user-defined AAL. The contents of the AAL parameter IE depends on the AAL type.

2.2.1.1 AAL Type 1

AAL type 1 is a connection-oriented, constant bit rate service with an end-to-end timing requirement. Examples of services that may use AAL type 1 include CBR audio, video, voice, and circuit emulation.

The AAL type 1 convergence sublayer (CS) may include various functions such as handling of cell delay variation, processing of the sequence count, providing a mechanism to transfer timing information, providing the transfer of structure information between source and destination, and forward error correction.

The AAL 1 segmentation and reassembly protocol data unit (SAR-PDU) header is illustrated in Figure 2.3. It is a 1-byte field carried in the cell payload. It is composed of a convergence sublayer indicator (CSI) bit, a 3-bit sequence number field, CRC-3, and a parity bit. The CSI bit carries the convergence sublayer indication. The CSI bit and the sequence number fields are protected by a 3-bit cyclic redundancy check code (CRC-3). The resulting 7-bit field is protected by an even parity check bit.

The 3-bit sequence number field provides sequence numbers from 0 to 7 (i.e., module 8) to detect lost cells.

In order to deliver user data at a constant bit rate, the convergence sublayer needs a clock. Three methods are allowed for the handling the timing relation between the called and the calling user: adaptive clock method, synchronous clock derived directly from the network clock, and synchronous residual time stamp method.

In the adaptive clock method, the receiver writes the received information into a buffer and reads it with a local clock. The buffer fill level is used to control the frequency of the local clock. The control is performed by continuously measuring the buffer fill level around its median position. If the fill level is

CSI	Sequence number	CRC-3	Even parity

Figure 2.3 AAL type 1 SAR-PDU header.

greater than the median, the local service clock is assumed to be slow and the local clock speed is increased. If the fill level is lower than the median, the clock is assumed to be fast and the local clock speed is decreased.

Another option is to phase lock the local clocks to the network clock. In this case, both local clocks are synchronized as each synchronizes its clock to the same network clock. However, it is not always practical and/or preferred to synchronize local clocks to the network clock. An example is the circuit emulation service used, for example, to provide an ATM network connectivity between two T1 services. The clocks of each service may not be synchronized to each other (i.e., two different network clocks) and the interworking unit between the two networks cannot synchronize its local clock to two different clocks simultaneously.

The synchronous residual time stamp (SRTS) method provides a measure of information about the frequency difference between a locally available reference clock derived from the network and the source clock. To synchronize the sender and receiver to the same frequency, the method assumes that the same network reference clock is available to both entities. The method determines the residual of the difference between the local service clock and the network clock and transmits it using the CSI bit of the SAR header in odd sequence number cells. Hence, eight ATM cells are required to pass one RTS value.

The 47-byte payload used by the CS has two formats: P and non-P. In the non-P format, the entire payload is used for user information with the first byte of the payload being the pointer field. This is referred to as structured data transfer. The P format is used only with even sequence numbers. The seven bits of the pointer field in the P-format are used to indicate the offset, measured in bytes, between the end of the pointer field and the first start of the structured block in the 93-byte payload consisting of the (current) 46-byte payload and the 47 bytes of the next CS-PDUs.

Another function currently defined for AAL 1 is the correction method for bit errors and cell losses for unidirectional video services. This method combines forward error correction (FEC) and octet interleaving. In the transmitting CS, a special 4-byte code (i.e., Reed-Solomon code) is appended to incoming data from the AAL user. Figure 2.4 illustrates the format of AAL type 1 IE.

The subtype field is used to specify the application type. Currently defined values are

- Null;
- Voice-band signal transport based on 64 Kbps;
- Circuit transport;
- High-quality audio signal transport;
- Video signal transport.

Field	Octets
Subtype identifier	6
Subtype	6.1
CBR rate identifier	7
CBR rate	7.1
Multiplier identifier	8*
Multiplier	8.1–8.2
Source clock frequency recovery method identifier	9*
Source clock frequency recovery method	9.1
Error-correction method identifier	10*
Error-correction method	10.1
Structured data transfer block size identifier	11*
Structured data transfer block size	11.1–11.2
Partially filled cells method identifier	12*
Partially filled cells method	12.1

Figure 2.4 The format of AAL type 1 IE.

The CBR rate specifies the transfer rate of the application. Currently allowed values are shown in Table 2.8.

Table 2.8
Current Allowed Transfer Rates

64 Kbps (DS1)
1,544 Kbps (T1)
6,312 Kbps (J2)
32,064 Kbps
44,736 Kbps (T3)
97,728 Kbps
2,048 Kbps (E1)
8,448 Kbps (E2)
34,368 Kbps (E3)
139,264 Kbps (E4)
$n \times 64$ Kbps
$n \times 8$ Kbps

For CBR rates of $n \times 64$ Kbps and $n \times 8$ Kbps, the multiplier field is used to specify the value of n. The 2-byte field is used as an integer multiplier value between 2 and $2^{16} - 1$ for $n \times 64$ Kbps and between 1 and 7 for $n \times 8$ Kbps.

The source clock frequency recovery method field allows the specification of three methods: *null* method (used for synchronous circuit transport), *SRTS* method (used for asynchronous circuit transport), and *adaptive clock* method.

The error-correcting method field allows the specification of null (no error correction is provided), a forward error-correction method for loss-sensitive signal transport, and a forward error-correction method for delay-sensitive signal transport. These functions are currently being studied in various standards bodies.

Structured data transfer (SDT) block size allowed in AAL 1 ranges between 1 and 65,535. When provisioning ATM connections that support AAL type 1 SDT service, the SDT protocol may distinguish between SDT block sizes with a value of 1 and others ranging from 2 to $2^{16} - 1$. The special case of using a block size of 1 is currently being studied in various standards bodies.

Not all 47 bytes of AAL 1 payload may be used by the application. For example, it takes about 6 ms to fill the cell with voice frames at the rate of 64 Kbps. If this delay is not tolerable, partially filled cells may be used to reduce the delay. The partially filled cells method allows the specification of the integer representation of the number of leading octets of AAL 1 payload in use (values between 1 and 47). However, the capability to support partially filled cells is not supported in the current version of Q.2931.

Subtype and CBR rate fields are mandatory fields in the AAL 1 information element. The default values of other fields are null for clock frequency, null for error correction, and no SDT is used for SDT block size. The multiplier field is mandatory for CBR rate $n \times 64$ Kbps and $n \times 8$ Kbps services.

2.2.1.2 AAL Type 3/4

AAL 3/4 is defined for connection-oriented and connectionless VBR services that do not require a timing relationship between the source and the destination. AAL 3/4 can be used to provide either assured operations or unassured operations. Every data unit in the assured mode is required to be delivered to the receiver correctly. This may require retransmission of missing or corrupted data units and that the protocol for the two end entities to exchange messages. Flow control is mandatory in assured operations. In the case of unassured operation, lost or corrupted SDUs are not corrected by retransmission. Both flow control and delivery of corrupted SDUs to the receiver are optional functions.

AAL 3/4 uses a 10-bit multiplexing identification field (MID) that can be used to multiplex two or more AAL connections into a single ATM connection. All data units of a particular AAL 3/4 connection are assigned the same MID value. As data units from different AAL 3/4 connections will have different MID values, it is possible to interleave and reassemble different AAL data units. A typical example on the use of multiplexing feature of AAL 3/4 and the MID field is the connectionless service in ATM networks where several CS-PDUs originally belonging to different sources may be transported using the same VPI/VCI. In this scenario, all packets originating at one LAN may use the same connection from the origin border node to a connectionless server. These cells are demultiplexed at the server based on their MID values. Figure 2.5 illustrates the format of AAL type 3/4 IE.

Both forward and backward maximum CPCS-SDU size is a 16-bit integer representing the range of values between 0 and 65,535. MID range fields includes the integer representation of the lowest MID value and the highest MID value, both between 0 and 1,023. Four SSCP types are currently specified: *null SSCS, assured operation with data SSCS based on SSCOP, unassured operation with data based on SSCOP*, and *frame relay SSCS*.

The default CPCS-SDU size (both forward and backward) is 65,535 octets. If the MID range fields are 0 and 0, the default is no multiplexing via the MID field. Finally, the default SSCS type is null.

2.2.1.3 AAL Type 5

AAL 5 is specified for VBR services that do not require a timing relation between the source and the destination and provide either assured operations or unassured operations. Figure 2.6 illustrates the format of AAL type 5 IE.

Similar to AAL 3/4, both forward and backward maximum CPCS-SDU size in AAL 5 is a 16-bit integer representing the range of values between 0 and 65,535 bytes. Four SSCP types are currently specified in AAL 5: *null SSCS, assured operation with data SSCS based on SSCOP, unassured operation with data based on SSCOP*, and *frame relay SSCS*.

The default value of CPCS-SDU size is 65,535 bytes and the default SSCS type is *null*.

Field	Octets
Forward maximum CPCS-SDU size identifier	6*
Forward maximum CPCS-SDU size	6.1–6.2
Backward maximum CPCS-SDU size identifier	7*
Backward maximum CPCS-SDU size	7.1–7.2
MID range identifier	8*
Lowest MID value	8.1–8.2
Highest MID value	8.3–8.4
SSCS type identifier	9*
SSCS type	9.1

Figure 2.5 The format of AAL type 3/4 IE.

Field	Octets
Forward maximum CPCS-SDU size identifier	6*
Forward maximum CPCS-SDU size	6.1–6.2
Backward maximum CPCS-SDU size identifier	7*
Backward maximum CPCS-SDU size	7.1–7.2
SSCS type identifier	8*
SSCS type	8.1

Figure 2.6 The format of AAL type 5 IE.

2.2.1.4 User-Defined AAL

User-defined AAL allows the users to define their own (proprietary) AAL. This capability in Q.2931 is provided for the calling user to pass proprietary AAL information to the called user. The contents of the user-defined AAL fields are user-specified, as shown in Figure 2.7.

2.2.2 Broadband High-Layer Information IE

The *broadband high-layer information* (B-HLI) IE is used for compatibility checking by an addressed entity. It is transferred transparently by the network between the calling user and the called user. The details of compatibility checking are discussed in Section 2.4.2.1.

The format of the broadband high-layer information IE is illustrated in Figure 2.8.

The high-layer information type could be *ISO/IEC* (International Standards Organization/inter exchange carrier), *user-specific, vendor-specific application identifier*, or *reference to B-ISDN teleservice recommendation*. The use of the

Field	Octets
User-defined AAL information	5.1–5.4*

Figure 2.7 The format of user-defined AAL.

Field		Octets
1 ext	High-layer information type	5
High-layer information element		6–13*

Figure 2.8 The format of the broadband high-layer information IE.

ISO/IEC code point is specified in the ISO/IEC standards. The coding of high-layer information IE for user-specific applications is user defined and requires an agreement between the two end users. If the vendor-specific application identifier is specified, octets 6–8 contain a globally administered, organizationally unique identifier (OUI) and octets 9–12 contain an application identifier administered by the vendor identified by the OUI. Octet 13 is not used in this case. In the case of B-ISDN teleservice recommendation, octet 6 indicates the specific code point defined for the service. The related code points will be added in the future when ITU-T completes the corresponding recommendations.

2.2.3 Broadband Low-Layer Information IE

The purpose of the *broadband low-layer information* (B-LLI) IE is to provide a means for compatibility checking by the called user. When the calling user wishes to notify the called user of its low-layer protocols above the ATM adaptation layer, it includes a B-LLI IE in the SETUP message. This IE is conveyed by the network and delivered to the called user. The B-LLI IE is transferred transparently by the network between the calling and the called user. The format of the B-LLI IE is illustrated in Figure 2.9.

When octet 6 indicates a user-specified layer 2 protocol, octet 6a specifies the actual user protocol. Octets 7a or 7a and 7b may be present to indicate protocols that do not have code points in octet 7. They can be user-defined protocols [e.g., Systems Network Architecture (SNA)] or protocols that are identified by TR 9577 [e.g., Internet Protocol (IP)]. Octet group 8 identifies

Field				Octets
0/1 ext	1	0	User information layer 2 protocol	6*
	Layer 2 identifier			
1 ext	User-specified layer 2 protocol information			6a*
0/1 ext	1	1	User information layer 3 protocol	7*
	Layer 3 identifier			
1 ext	User-specified layer 3 protocol information			7a*
0 ext	ISO/IEC TR 9577 Initial Protocol Identifier (IPI) (bits 8 to 2)			7a*
1 ext	IPI (bit 1)		Spare	7b*
1 ext	0	0	Spare	8*
	SNAP ID			
OUI				8.1–8.3*
PID				8.4–8.5*

Figure 2.9 The format of the broadband low-layer information IE.

the layer 3 protocol using SNAP identification (IEEE 802.1) and is included only when octet group 7 indicates TR 9577.

The user information protocol fields of the B-LLI IE indicate the low-layer protocols (i.e. layer 3 and layer 2 protocols above the AAL) used between endpoints (users). This information is not interpreted by the network and therefore the bearer capability provided by the B-ISDN is not affected by this information. The addressed entity may modify the low-layer attributes using the negotiation procedure described next. The negotiation procedures are optional in the network and may not be supported.

If the calling user wishes to indicate alternative values of B-LLI parameters (e.g., alternative protocol suites), the B-LLI IE is repeated in the SETUP message. Up to three B-LLI IEs may be included in a SETUP message. The first B-LLI IE in the message is preceded by the *broadband repeat indicator* IE specifying the *priority list for selecting one possibility (descending order of priority)*. That is, the order of appearance of the B-LLI IE indicates the order of preference of end-to-end low-layer parameters.

If the network or called user does not support multiple B-LLI IEs, it discards the broadband repeat indicator IE and the subsequent B-LLI IEs. Only the first B-LLI IE is used in the negotiation. In addition, if the network discards the B-LLI IE, it sends a STATUS message with the cause "access information discarded."

The called user indicates a single choice from among the options offered in the SETUP message by including the B-LLI IE in the CONNECT message. Absence of a B-LLI IE in the CONNECT message indicates acceptance of the first B-LLI IE in the SETUP message. If the calling user rejects the B-LLI IE contents in the CONNECT message, the calling user initiates call clearing with cause "invalid information element contents."

2.2.4 Called Party Number Information Element

The *called party number* IE is used to identify the called party of a call. The format of this IE is illustrated in Figure 2.10.

Field			Octets
1 ext	Type of number	Addressing/numbering plan identification	5
0	Address/number digits		6 etc. or
NSAP address octets			6 etc.

Figure 2.10 The format of the called party number IE.

The most commonly used combinations of addressing/numbering plans and types of numbers are

- Unknown—ATM endsystem address (ISO NSAP);
- International number—ISDN/telephony numbering plan (E.164);
- National number—ISDN/telephony numbering plan (E.164).

The complete ATM address requires multiple occurrences of octet 6 with bit 8 set to 0 in each unless it is an NSAP address, in which case all 8 bits of the octet are used in each occurrence of octet 6.

2.2.5 Called Party Subaddress Information Element

The *called party subaddress* IE is used to identify the subaddress of the called party of a call. Its format is illustrated in Figure 2.11.

The type of subaddress can be *NSAP, user-specified ATM endsystem address*, or *user-specified*. The user-specified ATM endsystem address can be used by private networks to carry ATM endsystem addresses across the public network. The content of the subaddress information field depends on the type of the subaddress field. In the case of a user-specified subaddress, this field is encoded according to the user specification, subject to a maximum length of 20 octets. An odd/even indicator field is used to indicate either an even or odd number of address digits. When the odd/even indicator is set to *odd number of digits,* the high-order four bits are not used (i.e., it does not contain an address digit). This indicator is used when the type of subaddress is user specified and the coding is BCD (binary coded decimal).

2.2.6 Calling Party Number Information Element

The *calling party number* IE is used to identify the origin of a call. Its format is illustrated in Figure 2.12.

Field						Octet
1 ext	Type of subaddress	Odd/ even indicator	0	0 Spare	0	5
Subaddress information						6 etc.

Figure 2.11 Called party subaddress IE format.

Field								Octet
8	7	6	5	4	3	2	1	*Octet*
Information element identifier								1
1 ext	Coding standard	IE instruction field Flag	Reserved	IE action indicator field				2
Length of information element parameter contents								3–4
Information element specific information								5–etc.

Figure 2.12 Calling party number IE format.

The calling party number IE fields are the same as in the called party number IE except for the use of the following two fields: *presentation indicator* and *screening indicator*. At the originating UNI, the presentation indicator is used to indicate the intention of the calling user for the presentation of the calling party number to the called user. This may also be requested on a subscription basis. The 2-bit presentation indicator field values are *presentation allowed, presentation restricted,* and *number not available.* If octet 5a is omitted and the network does not support subscription information for the calling party number information restrictions, the *presentation allowed* is assumed.

The screening indicator field values are *user-provided not screened, user-provided verified and passed, user-provided verified and failed,* and *network-provided.*

An interested reader may refer to Chapter 5 for the details of presentation allowed and screening indicators used for various supplementary services.

2.2.7 Calling Party Subaddress Information Element

The *calling party subaddress* IE is used to identify a subaddress associated with the origin of a call. The format of this information element is exactly the same as the called party subaddress IE.

2.2.8 Cause Information Element

The *cause* IE describes the reason for generating certain messages and indicates the location of the cause originator. The cause IE may be repeated in a message. The contents and the use of the cause IE are defined in Q.2610 and described throughout the book in corresponding sections. Its format is illustrated in Figure 2.13.

The cause value field indicates the reason for generating the message. The cause values are specified in ATM Forum UNI 3.1 and ITU Recommendation

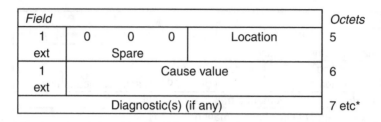

Field					Octets
1 ext	0	0 Spare	0	Location	5
1 ext	Cause value				6
Diagnostic(s) (if any)					7 etc*

Figure 2.13 Cause IE format.

Q.2610. Some cause values require additional diagnostic information. For example, for cause "information element nonexistent or not implemented," the diagnostic field includes the information element identifier(s).

2.2.9 Connection Identifier Information Element

The *connection identifier* IE is used to identify the VPI/VCI allocated to the connection at an interface. This IE is optionally included in the SETUP message in the user-to-network direction to request a particular connection identifier from the network. It is mandatory in the first response to the SETUP message in network-to-user direction. The format of this IE is illustrated in Figure 2.14.

As discussed in Chapter 1, two modes of signaling are supported at the UNI: (1) associated signaling in which the layer 3 signaling entity exclusively controls the VCs in the VPC that carries its signaling VC and (2) nonassociated signaling in which the layer 3 signaling entity controls the VCs in the VPC that carry its signaling VC and VCs in other VPCs that are configured to be controlled by the signaling VC. The network and user are required to support nonassociated signaling and may, as an option, support associated signaling procedures. The associated signaling procedures are used only by bilateral agreement between the user and the network.

The connection identifier IE contains the VPCI instead of VPI, since, in general, virtual path cross connects may be used in the access and multiple inter-

Field					Octets
1 ext	0	0 Spare	VP-assoc. signaling	Preferred/exclusive	5
VPCI					6–7
VCI					8–9

Figure 2.14 Connection identifier IE format.

faces could be controlled by the signaling virtual channel. The ATM Forum UNI 3.1 specification supports only single interface with no cross connects. In this case, the VPCI value is same as the VPI value.

In associated signaling, the user requests a VC in the VPC to carry the signaling messages (i.e., a signaling VC). The VPC carrying the signaling VC is implicitly indicated. In the connection identifier IE, the VP-associated signaling field is coded as *VP-associated signaling* and one of the following values is indicated in the preferred/exclusive field: exclusive VPCI, any VCI; or exclusive VPCI, exclusive VCI.

In nonassociated signaling, the user's request for a VC in the connection setup message includes one of the following: exclusive VPCI, any VCI; exclusive VPCI, exclusive VCI; or no indication is included (i.e., the connection identifier IE is not included in the SETUP message).

The VPI field indicates the identifier of the virtual path connection. A VPI is an integer between 0 and 255 (at the UNI). The exact range of VPI values supported in the network is determined during the subscription time. In any VPI, VCI values 0 to 31 are reserved by ITU-T and ATM Forum and cannot be used for switched user connections. The range of VCI values available for switched connections is between 32 and 65,535. However, the whole range may not be available to users and the network may restrict the number of bits supported in the VCI field. The exact range of VCI values supported in the network is determined during the subscription time.

2.2.10 End-to-End Transit Delay Information Element

The *end-to-end transit delay* IE is used to indicate (1) the maximum, one-way, end-to-end transit delay between the calling user and the called user acceptable on a per-call basis and (2) the cumulative transit delay to be expected for a virtual channel connection. It includes the total processing time in the end-user systems (e.g., processing time, AAL handling delay, ATM cell assembly delay, etc.) and the network transfer delay (e.g., propagation delay, ATM layer transfer delay, possibly any additional processing delay in the network). The format of the end-to-end transit delay IE is illustrated in Figure 2.15.

Field	Octets
Cumulative transit delay identifier	5
Cumulative transit delay value	5.1–5.2
Maximum end-to-end transit delay identifier	6*
Maximum end-to-end transit delay value	6.1–6.2

Figure 2.15 The format of the end-to-end transit delay IE.

The cumulative transit delay value indicated by the calling user in the SETUP message (if the IE is present) includes the cumulative transit delay from the calling user to the network boundary. The value indicated by the network in the SETUP message sent to the called user is the sum of the value that was indicated at the originating UNI and the expected transfer delay accumulated within the network. It does not include the additional transfer delay on the way from the network boundary to the called user. This is because the end station may not be directly attached to the UNI (i.e., the called end station may be an interworking unit and the user may be off to the unit). It is up to the called user to update the total delay value received in the SETUP message to include the delay from the network to the end station and the internal processing delays. The cumulative transit delay value transferred over both UNIs in the CONNECT message is the expected total end-to-end transit delay value for user data transfer over the corresponding virtual channel connection. The cumulative transit delay value is encoded in milliseconds.

The maximum end-to-end transit delay value may be indicated by the calling user to specify end-to-end transit delay requirements for this call. Inclusion of this field in the SETUP message indicates that the calling user has specified end-to-end transit delay requirements for this call. The maximum end-to-end delay value is encoded in milliseconds. The value 1111 1111 1111 1111 indicates that any end-to-end transit delay value acceptable but to deliver cumulative end-to-end transit delay value to the called user. The maximum end-to-end transit delay is not included in the CONNECT message.

2.2.11 Quality of Service Parameter Information Element

The *quality of service parameter* IE is used to indicate the QoS class requested for the connection. Its format is illustrated in Figure 2.16.

Three network-specific QoS classes are defined by ATM Forum:

- QoS class 1—Supports constant bit rate traffic with stringent timing requirements, such as circuit emulation;

Field	Octets
QoS class forward	5
QoS class backward	6

Figure 2.16 Quality of service parameter IE format.

- QoS class 2—Supports variable bit rate traffic with stringent timing requirements, such as compressed voice and video;
- QoS class 3—Supports connection-oriented, variable bit rate traffic with no timing requirements, such as frame relay.

In addition, both ATM Forum UNI and ITU Q.2931 signaling support QoS class 0: *Unspecified QoS class*. It is proposed to use class 0 as the default QoS. ITU I.356 specifies four additional QoS classes that will be supported in future versions of ITU signaling specifications.

2.2.12 Broadband Repeat Indicator Information Element

The *broadband repeat indicator* IE is used to indicate how repeated IEs are interpreted. It is included before the first occurrence of the IE that will be repeated in a message. The format of this IE is illustrated in Figure 2.17.

Descending order of priority is the only priority mechanism supported in Q.2931. That is, the first IE has the highest priority, the second IE has the second highest priority, and so forth.

2.2.13 Transit Network Selection Information Element

The *transit network selection* IE is used to identify a transit network that the call may cross. This IE may be repeated in a message to select a sequence of transit networks through which a call is requested to pass through. Its format is illustrated in Figure 2.18.

Values defined for the type of network identification field are *national network identification* and *international network identification*. Similarly, the net-

Field					Octets
1 ext.	0	0 Spare	0	Broadband repeat indication	5

Figure 2.17 The format of the broadband repeat indicator IE.

Field			Octets
1 ext.	Type of network identification	Network identification plan	5
0	Network identification		6 etc.

Figure 2.18 Transit network selection IE format.

work identification plan field values are *carrier identification code* and *data network identification code* (X.121). Finally, the network identification field includes the characters organized according to the network identification plan.

2.3 Call and Connection Control Messages

Each Q.2931 signaling message uses the message format defined in Chapter 1. The first nine octets contain the common format followed by one or more IEs, starting at octet 10. The protocol discriminator used by Q.2931 is 00001001. ATM equipment (user or network) may not be able to process or understand every IE included in a message. If an IE is not needed for its proper operation, the equipment normally ignores that information. For example, a user device may ignore the calling party number if that number is of no interest to the user. Similarly, the network may not process the ATM adaptation layer parameters.

Table 2.9 summarizes the messages used by Q.2931 call/connection control. The details of each message is described in subsequent sections.

Table 2.9
Call Control Messages

Message Type	Significance	Direction	Description
ALERTING	Global	Both	Indicates that the called user alerting has been initiated
CALL PROCEEDING	Local	Both	Indicates that the requested call/connection establishment has been initiated and no more call establishment information will be accepted
CONNECT	Global	Both	Indicates that the call/connection request is accepted by the called user
CONNECT ACKNOWLEDGE	Local	Both	In the network (user)-to-user (network) direction, it indicates that the network (user) has accepted the call
SETUP	Global	Both	Initiates a call/connection establishment

2.3.1 ALERTING

The ALERTING message is sent by the called user to the network and by the network to the calling user to indicate that the called user alerting has been initiated (see Figure 2.19). This message is used for calls that use human interface (e.g., a voice call).

The connection identifier IE is included if the ALERTING message is the first response to a SETUP message.

2.3.2 CALL PROCEEDING

The CALL PROCEEDING message is sent by the called user to the network or by the network to the calling user to indicate that the requested call establishment has been initiated and no more call establishment information will be accepted (see Figure 2.20).

Symmetrical procedures are optional procedures that enable the protocol to be the same at the originating and terminating interfaces. These procedures are required when Q.2931 is used between two peer entities connected directly without a network between them, for example when two end stations are attached to each other in a point-to-point manner.

When symmetrical procedures are not supported, the CALL PROCEEDING message is the first response in the network-to-user direction. The connection identifier IE is mandatory in the network-to-user direction if it is the first response to a SETUP message. It is also mandatory in the user-to-network direction if this message is the first message in response to a SETUP message, unless the user accepts the connection identifier indicated in the SETUP message. That is, if the called party accepts the connection identifier in the SETUP message, it does not send the connection identifier in its CALL PROCEEDING message.

Information Element	Type	Length
Connection identifier	O	4–9

Figure 2.19 Information elements used in ALERT message.

Information Element	Type	Length
Connection identifier	O	4–9

Figure 2.20 Information elements used in CALL PROCEEDING message.

2.3.3 CONNECT

The CONNECT message is sent by the called user to the network and by the network to the calling user to indicate call acceptance by the called user (see Figure 2.21).

An AAL parameter IE is included in the user-to-network direction when the called user wants to pass final parameters used after AAL negotiation to the calling user.

A broadband low-layer IE is included to support B-LLI negotiation and to indicate the values of parameters after the negotiation.

A connection identifier IE is mandatory in the user-to-network direction if the message is the first one in response to a SETUP message, unless the user accepts the connection identifier indicated in the SETUP message.

An end-to-end transit delay IE indicates the delay expected for user information transfer. It is included in the CONNECT message when the responding user receives the end-to-end transit delay IE in the SETUP message.

2.3.4 CONNECT ACKNOWLEDGE

This message is sent by the network to the called user to indicate the user has been awarded the call. It is also sent by the calling user to the network to allow symmetrical call control procedures.

2.3.5 SETUP

This message is sent by the calling user to the network and by the network to the called user to initiate B-ISDN call and connection establishment (see Figure 2.22).

An AAL parameter IE is included when the calling user wants to pass AAL information to the called user. Broadband high-layer information, broadband low-layer information, called party subaddress, and calling party subaddress IEs are included to check compatibility with the called user (refer to Section 2.4.2.1 for the details).

Information Element	Type	Length
AAL parameters	O	4–11
Broadband low-layer information	O	4–17
Connection identifier	O	4–9
End-to-end transit delay	O	4–7

Figure 2.21 Information elements used in CONNECT message.

Information Element	Type	Length
AAL parameters	O	4–21
ATM traffic descriptor	M	12–30
Broadband bearer capability	M	6–7
Broadband high-layer information	O	4–13
Broadband repeat indicator	O	4–5
Broadband low-layer information	O	4–17
Called party number	O	4–25
Called party subaddress	O	4–25
Calling party number	O	4–26
Calling party subaddress	O	4–25
Connection identifier	O	4–9
End-to-end transit delay	O	4–10
QoS parameter	M	6
Transit network selection	O	4–9

Figure 2.22 Information elements used in SETUP message.

A broadband repeat indicator IE is included when two or more broadband low-layer information IEs are included for low-layer information negotiation. It precedes the first broadband low-layer information IE. Two or three IEs may be included in descending order of priority (i.e., highest priority first) if the broadband low-layer information negotiation procedures are used.

A called party IE is included by the user to convey called party number information to the network. It is mandatory in the user-to-network direction.

A calling party number IE may be included by the calling user or the network to identify the calling user.

A connection identifier IE is included in the user-to-network direction when a user wants to indicate a virtual channel. If it is not included, its absence is interpreted as meaning that any virtual channel is acceptable. It is included in the network-to-user direction when the network wants to indicate the allocated virtual channel.

An end-to-end transit delay IE is included when the calling user wants to specify the end-to-end transit delay requirements for its call and/or the cumulative transit delay expected for the transmission of user data from the calling user to the network boundary.

A transit network selection IE is included by the calling user to select a particular transit network. This IE may appear up to four times in the SETUP message.

2.3.6 Coding Rules for Information Elements

The rules for coding the variable length IEs are formulated to allow each piece of equipment to find the IEs it needs to process in a message and yet remain ignorant of IEs that are not used by it. For example, an IE may be present, but empty. Or, a SETUP message may contain a null calling party number IE (i.e., with a zero length content). This is interpreted by the receiver as if the IE is not included. IEs within a message may appear in any order except for the following:

1. If IEs are repeated without using the broadband repeat indicator IE, the repeated IEs need to be consecutive in the message.
2. When IEs are repeated and the broadband repeat indicator IE is used, the following rules apply:

   - The broadband repeat indicator IE must immediately precede the first occurrence of the repeated IE.
   - The second occurrence of the repeated IE must immediately follow the first occurrence of the repeated IE. The third occurrence of the repeated IE must immediately follow the second occurrence of the repeated IE, and so forth.

2.3.6.1 Structure Using Information Subfield Identifiers

Each octet or octet group is numbered to facilitate referencing. The first digit in the octet number identifies one octet or a group of octets. The octet groups are coded using a subfield identifier or an extension mechanism.

Using Subfield Identifiers

The subfield identifiers are position independent (i.e., they need not appear in a certain order within the information element). If an octet group is present, all the octets of that group are included. Each octet group is a self-contained entity.

Optional octet groups are marked with asterisks (*). The subfield identifier implicitly defines the contents (i.e., number of octets). An example of the structure of a Q.2931 IE using the subfield identifiers is shown in Figure 2.23.

Using the Extension Mechanism

The extension mechanism is used to extend an octet (N) through the next octet(s) (Na, Nb, etc.) by using bit 8 in each octet as an extension bit. The bit value of 0 indicates that the octet continues through the next octet, whereas a value of 1 indicates that this octet is the last octet. If octet (Nb) is present, the preceding octets (N and Na) must be present as well. Optional octets are marked with asterisks (*). This case is illustrated in Figure 2.24, and an example is given in Figure 2.25.

In addition to the extension mechanism, an octet can be extended without using an extension bit. This mechanism is used only at the end of the IE, in which case the octet group length can be derived from the IE length.

Field	Octets
Cumulative transit delay identifier	5
Cumulative transit delay value	5.1–5.2
Maximum end-to-end transit delay identifier	6*
Maximum end-to-end transit delay value	6.1–6.2

Figure 2.23 Example of subfield identifiers—end-to-end transit delay IE.

Bit 8	Meaning
0/1 ext.	Another octet of this octet group may follow.
1 ext.	This is the last octet in the extension domain.
0 ext.	Another octet of this octet group always follows.

Figure 2.24 Extensions of octets in an IE.

0/1 ext	Type of number	Addressing/numbering plan identification					5
1 ext	Presentation indicator	0	0	0		Screening indicator	5a*
0		Address/number digits					6 etc or
		NSAP address octets					6 etc

Figure 2.25 Example of extension bit—calling party number IE.

The mechanisms of subfield identifiers and extension bits may be combined. An example of the structure of a Q.2931 IE using both subfield identifiers and an extension bit is shown in Figure 2.26.

In Figure 2.26, octet groups 6 and 7 use the extension mechanism. In octet group 8, the extension mechanism is not used (in octets 8.1 to 8.5). Hence, the octet group 8 can only be the last octet group.

2.3.6.2 Extensions of Code Sets

It is possible to expand this structure to eight code sets. One common value of an IE identifier is employed in each code set to facilitate shifting from one code set to another. The contents of this *shift* IE identifies the code set to be used for the next set of IE(s). The code set in use at any given time is referred to as the *active code set.* By convention, code set 0 is the initially active code set. Two code set shifting procedures are supported: locking shift and nonlocking shift (see Table 2.10).

Table 2.10

ITU-T Specified Code Sets

Code Set	Reserved for
1 to 3	Future ITU use
4	Use by ISO/IEC standards
5	IEs for national use
6	IEs specific to the local network (either public or private)
7	user-specific IEs

Field				Octets
0/1 ext	1 Layer 2 identifier	0	User information layer 2 protocol	6*
1 ext	User-specified layer 2 protocol information			6a*
0/1 ext	1 Layer 3 identifier	1	User information layer 3 protocol	7*
1 ext	User-specified layer 3 protocol information			7a*
0 ext	ISO/IEC TR 9577 Initial Protocol Identifier (IPI) (bits 8 to 2)			7a*
1 ext	IPI (bit 1)	Spare		7b*
1 ext	0 SNAP ID	0	Spare	8*
	OUI			8.1–8.3*
	PID			8.4–8.5*

Figure 2.26 Example of combined mechanisms—broadband low-layer information IE.

2.4 Call/Connection-Establishment Procedures

Before the Q.2931 procedures are invoked, an assured mode signaling AAL connection must be established between the user and the network. All layer 3 messages are sent to the signaling AAL using an AAL-DATA-request primitive.

2.4.1 Connection Establishment at the Originating Interface

2.4.1.1 Call/Connection Request

A new call/connection establishment is initiated by transferring a SETUP message on the assigned signaling virtual channel across the interface. Upon sending a SETUP message, the calling party starts timer T303 and the call enters the *call initiated* state. The message contains a unique call reference value at the interface. The ATM traffic descriptor, broadband bearer capability, and QoS parameter IEs are mandatory in the SETUP message. Since *overlap* sending is not supported, the SETUP message also contains the called party address information in the called party number IE and possibly the called party subaddress IE.

If no response to the SETUP message is received by the user before timer T303 expires for the first time, the SETUP message is retransmitted and timer T303 is restarted. If the user has not received any response to the SETUP message after the second time the timer expires, the calling user clears the call internally.

As discussed in Chapter 1, the connection identifier allocation/selection procedure at the originating interface depends on whether associated signaling or nonassociated signaling is used.

In associated signaling, the user requests a VC in the VPC carrying the signaling VC. The VPC carrying the signaling VC is implicitly indicated. In the connection identifier IE, the VP-associated signaling field is coded as *VP-associated signaling* and one of the following values is used in the preferred/exclusive field: exclusive VPCI, any VCI; or exclusive VPCI, exclusive VCI. In the former case, the network selects any available VCI within the VPC carrying the signaling VC. If no VCI is available, the connection-clearing procedures are initiated. In the latter case, if the indicated VCI within the VPC carrying the signaling VC is available, the network selects it for the call. If the indicated VCI is not available, the connection-clearing procedures are initiated.

In nonassociated signaling, one of the following choices is indicated: (1) exclusive VPCI, any VCI; (2) exclusive VPCI, exclusive VCI; or (3) no indication is included.

In the case of (1), if the indicated VPCI is available, the network selects any available VCI in the VPCI for the call. If the specified VPCI is not available, a RELEASE COMPLETE message with cause "requested VPCI/VCI not available" is sent by the network. If no VCI is available, a RELEASE COMPLETE message with cause "no VPCI/VCI available" is sent by the network. If the VPCI value in the first response message is not the VPCI value indicated by the user, a RELEASE message with cause "VPCI/VCI assignment failure" is sent to the network.

In the case of (2), if the indicated VPCI and the VCI are both available within the VPCI, the network selects it for the call. If the specified VPCI is not available, a RELEASE COMPLETE message with cause "requested VPCI/VCI not available" is sent by the network. If the VCI in the indicated VPCI is not available, a RELEASE COMPLETE message with cause "requested VPCI/VCI not available" is sent by the network. If the VPCI and VCI values in the first response message are not the VPCI and VCI values indicated by the user, a RELEASE message with cause "VPCI/VCI assignment failure" is sent to the network.

In the case of (3), the network selects any available VPCI and VCI. If the network is not able to allocate a VCI in any VPCI, a RELEASE COMPLETE message with cause "no VPCI/VCI available" is sent by the network.

In all cases, the selected VPCI/VCI value is indicated in the connection identifier IE in the first message returned by the network in response to the SETUP message.

2.4.1.2 QoS and Traffic Parameter Selection Procedures

The user indicates the QoS class in the *quality of service parameter* IE. If the network is able to provide the requested QoS class, it progresses the call to the called user. Otherwise, the network rejects the call by returning a RELEASE COMPLETE message with cause "quality of service unavailable."

The user is required to indicate the requested traffic parameters in the ATM traffic descriptor IE. If the network is able to provide the requested traffic parameters, it progresses the call to the called user. Otherwise, the network rejects the call by returning a RELEASE COMPLETE message with cause "user cell rate unavailable."

2.4.1.3 Invalid Call/Connection Control Information

If upon receiving the SETUP message, the network determines that the call information received from the user is invalid (e.g., invalid number), it initiates call-clearing procedures with one of the following causes, as appropriate:

- Unassigned (unallocated) number;
- No route to destination;
- Number changed;

- Invalid number format (address incomplete).

While progressing a call, the network checks the valid combinations of traffic- and QoS-related parameters for the requested bearer service in the SETUP message as described in the subsequent sections.

2.4.1.4 Call/Connection Proceeding

If the network can determine that access to the requested service is authorized and available, it sends a CALL PROCEEDING message to the user to acknowledge the SETUP message and to indicate that the call is being processed. Upon sending the CALL PROCEEDING message, it enters the *outgoing call proceeding* state. When the user receives the CALL PROCEEDING message, it stops timer T303, starts timer T310, and enters the outgoing call proceeding state.

If the network determines that a requested service is not authorized or is not available, it initiates call clearing with one of the following causes, as appropriate:

- Bearer capability not authorized;
- Bearer capability not presently available;
- Service or option not available, unspecified;
- Bearer service not implemented.

If the user has received a CALL PROCEEDING message but does not receive an ALERTING, CONNECT, or RELEASE message prior to the expiration of timer T310, it initiates clearing procedures towards the network with cause "recovery on timer expiry."

2.4.1.5 Call/Connection Confirmation Indication

Upon receiving an indication that user alerting has been initiated at the called end station, the network sends an ALERTING message across the UNI of the calling end station and enters the *call delivered* state. When the user receives the ALERTING message, it may begin an internally generated alerting indication, stop timer T310, and enter the call delivered state.

2.4.1.6 Call/Connection Acceptance

Upon receiving an indication that the call has been accepted, the network sends a CONNECT message across the user-network interface to the calling user and enters the *active* state. This message indicates to the calling user that a connection has been established through the network.

Upon receiving the CONNECT message, the calling user stops timer T310 (if running) or timer T301, stops any user-generated alerting indication, attaches to the user plane virtual channel if not already done, sends a CONNECT ACKNOWLEDGE message, and enters the active state. At this point, an end-to-end connection is established.

The network does not take any action upon receiving a CONNECT ACKNOWLEDGE message when it perceives the call is in the active state.

The sequence of events that takes place at the user and the network side of the UNI from which the call originated are summarized in Figures 2.27 and 2.28, respectively.

2.4.1.7 Call/Connection Rejection

Upon receiving an indication that the network or the called user is unable to accept the call, the network initiates call clearing at the originating UNI,

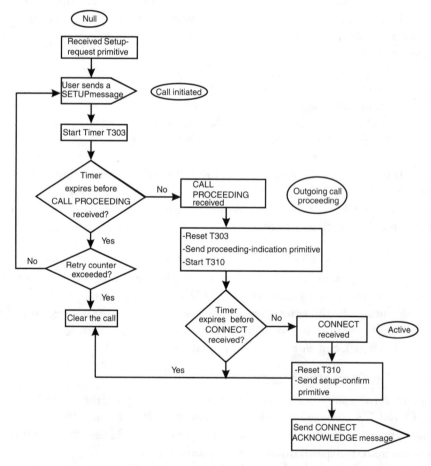

Figure 2.27 Connection establishment at the originating interface—user side without ALERTING message.

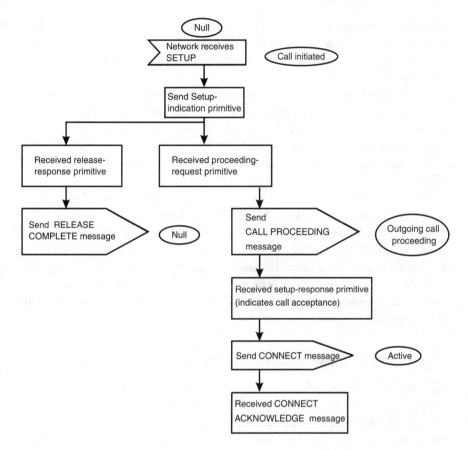

Figure 2.28 Connection establishment at the originating interface—network side without ALERTING message.

using the cause provided by the terminating network or the called user (cf. Section 2.5.2).

2.4.1.8 Transit Network Selection

If transit network selection is supported, the user identifies the selected transit network(s) in the SETUP message and uses a transit network selection IE to convey a single network identification. The user may specify up to four transit networks. When multiple transit networks are used, each selection is placed in a separate IE. The call would then be routed through the specified transit network(s) in the order listed in the SETUP message. Unlike Q.2931, UNI 3.1 and 4.0 allow the specification of a single transit network only.

As the call is delivered to each selected network, the corresponding transit selection may be stripped from the call establishment signaling, in accordance with the relevant inter-network signaling arrangement. The transit network selection IE(s) are not delivered to the called user. If a network cannot route the call due to insufficient bandwidth, it initiates call clearing with cause "user cell rate unavailable."

A network may screen the transit network selection IEs to (1) avoid routing loops, (2) ensure an appropriate business relationship exists between selected networks, or (3) ensure compliance with national and local regulations. If the transit network selection is of an incorrect format or fails to meet one of these criteria, the network initiates call clearing with cause "invalid transit network selection."

When a user includes the transit network selection IE, presubscribed default transit network selection (if any) is overridden. Some networks may not support transit network selection. In this case, when a transit network selection IE is received, it is processed according to the rules defined for unimplemented mandatory IEs by these networks, and the network initiates call clearing with cause "no route to specified transit network." The diagnostic field contains a copy of the contents of the transit network selection IE identifying the unreachable network.

2.4.2 Call/Connection Establishment at the Destination Interface

The network indicates the arrival of a call at the destination UNI the called party is attached to by transferring a SETUP message across the interface. Each message of a call contains the same unique call reference value to identify the call at the interface. Upon sending the SETUP message, the network starts timer T303 and enters the *call present* state. This message is sent by the network at the destination UNI only if the call is accepted by the network (i.e., required resources for the call are available). Upon receiving the SETUP message, the called user enters the call present state. If no response is received by the network to the first SETUP message before the timer T303 expires, the network retransmits the SETUP message and restarts timer T303. If timer T303 expires for the second time, the network starts clearing the call.

2.4.2.1 Compatibility Checking

The called user performs the compatibility check based on the compatibility information received in the SETUP message. In B-ISDN, there are two categories of compatibility information:

1. Broadband category 1 compatibility information is provided for both the network and the user to determine the attributes of the ATM con-

nection. Broadband category 1 compatibility information is always checked by the called user. If the compatibility check fails, the user is incompatible.

- Broadband bearer capability information;
- End-to-end transit delay information;
- ATM traffic descriptor;
- Quality of service parameter.

2. Broadband category 2 compatibility information is provided for the called user. Broadband category 2 compatibility information is always checked by the called user. If the compatibility check fails, the user is incompatible.

- ATM adaptation layer parameter information (describing the user plane AAL);
- Optional broadband low-layer information;
- Optional broadband high-layer information;
- Optional subaddress.

At the called user side, compatibility and address checking means that the user examines the contents of the specified IE. If an incoming SETUP message is offered with addressing information, the following actions take place: If a number or subaddress is assigned to a user, the address of the incoming call is checked by the user against the corresponding part of the number assigned to the user or the user's subaddress. In case of a mismatch, the user rejects the call. If a user has no assigned number or subaddress, then the address is ignored.

The user checks that the bearer service offered by the network in the broadband-bearer-related IEs match the bearer services that the user is able to support. If a mismatch is detected, the user rejects the offered call. When the user is unable to provide the requested QoS or to support the indicated traffic parameters, the call is rejected as well.

In summary, the user examines the following IEs upon receiving a SETUP message: broadband bearer capability, quality of service parameter, ATM traffic descriptor, and the end-to-end transit delay IEs (if present).

Next, the called side terminal equipment performs the following checks:

- Whether or not it can support the requested AAL type and its parameters specified in the AAL parameters IE;
- Whether or not the broadband low-layer information IE is compatible with the functions it supports. The B-LLI IE is used to check compatibility of low layers (e.g., layer 2 and layer 3);

- The called terminal equipment may check the broadband high-layer information IE as part of user-to-user compatibility checking procedures.

Upon agreement with other users or in accordance with other standards, the user-to-user IE may be used for additional compatibility checking.

If a mismatch is detected in checking any of the IEs above, the end station either ignores the IE or rejects the offered call.

2.4.2.2 Called User Clearing During Call Establishment

An incompatible user responds with a RELEASE COMPLETE message with cause "incompatible destination" and enters the *null* state. A busy user that satisfies the compatibility requirements indicated in the SETUP message normally responds with a RELEASE COMPLETE message with a cause "user busy." If the called user wishes to refuse the call, a RELEASE COMPLETE message is sent with cause "call rejected" and returns to the null state.

If a RELEASE COMPLETE or RELEASE message is received prior to receiving a CONNECT message, the network stops timer T303, timer T310, or timer T301 (if running), continues to clear the call to the called user, and clears the call to the calling user with the cause received in the RELEASE COMPLETE or RELEASE message.

2.4.2.3 Connection Identifier, QoS, and Traffic Parameter Selection

The connection identifier, QoS class, cumulative end-to-end transit delay traffic descriptor procedures at the destination UNI are the same as they are in the originating UNI. For connection identifier selection, if the network includes exclusive VPCI, exclusive VCI and the connection identifier IE is not present in the first response message, the connection identifier in the SETUP message is accepted.

2.4.2.4 Response to SETUP

When the called user determines that sufficient call setup information has been received and compatibility requirements are satisfied, it responds with either a CALL PROCEEDING, ALERTING, or CONNECT message and enters the incoming *call proceeding, call received,* or *connect request* state, respectively. The CALL PROCEEDING message is sent by the user if it cannot respond to a SETUP message with an ALERTING or CONNECT message before the expiry of timer T303.

Upon receiving a CALL PROCEEDING message from a called user, the network stops timer T303, starts timer T310, and enters the *incoming call proceeding* state.

Similarly, upon receiving an ALERTING message from a called user, the network stops timers T303 or T310 (if running), starts T301, enters the call received state, and sends a corresponding ALERTING message to the calling user.

If the network has received a CALL PROCEEDING message, but does not receive an ALERTING, CONNECT, or RELEASE message prior to the expiration of timer T310, it initiates clearing procedures towards the calling user with cause "no user responding" and initiates clearing procedures towards the called user with cause "recovery on timer expiry."

If the network has received an ALERTING message, but does not receive a CONNECT or RELEASE message prior to the expiration of timer T301, it initiates clearing procedures towards the calling user with cause "no answer from user (user alerted)" and initiates clearing procedures towards the called user with cause "recovery on timer expiry."

If a call can be accepted and no user alerting is required, a CONNECT message is sent without a previous ALERTING message.

2.4.2.5 Call/Connection Acceptance

A called user indicates acceptance of an incoming call by sending a CONNECT message to the network. Upon sending the CONNECT message, it starts timer T313 and enters the connect request state. If an ALERTING message had previously been sent to the network, the CONNECT message may contain only the call reference.

Upon receiving a CONNECT message, the network stops (if running) timers T301, T303, and T310, enters the connect request state, sends a CONNECT ACKNOWLEDGE message to the user, initiates procedures to send a CONNECT message towards the calling user, and enters the active state.

The CONNECT ACKNOWLEDGE message indicates the completion of the connection-establishment procedures. There is no guarantee that an end-to-end connection is established until a CONNECT message is received at the calling user. Upon receipt of the CONNECT ACKNOWLEDGE message, the called user stops timer T313, attaches to the user plane virtual channel, and enters the active state. If timer T313 expires prior to receipt of a CONNECT ACKNOWLEDGE message, the called user initiates clearing with cause "recovery on timer expiry."

The sequence of events that take place at the user and the network sides of a UNI the called end station is attached to are illustrated in Figures 2.29 and 2.30, respectively.

2.4.3 Call/Connection Establishment—State Diagrams

State transitions that take place at the user and the network side of a UNI are illustrated in Figures 2.31 and 2.32, respectively. Each transition indicates the input or output messages and primitives. The input to each state is in *italics*.

2.4.4 ATM Adaptation Layer Parameters Indication in the SETUP Message

When the calling endpoint wishes to indicate to the called endpoint the AAL common part parameters and service-specific part to be used during the call, the calling endpoint includes an ATM adaption layer parameters IE in the SETUP message. This information element is conveyed by the network and delivered to the called user.

2.4.4.1 Maximum CPCS-SDU Size Negotiation

When the called user has received an ATM adaptation layer parameters IE in a SETUP message and the AAL type is either AAL 3/4 or AAL 5, it includes the ATM adaptation layer parameters IE in the CONNECT message. The forward maximum CPCS-SDU size field in this IE indicates the size of the largest CPCS-SDU that the called user is able to receive. Similarly, the backward maximum CPCS-SDU size indicates the size of the largest CPCS-SDU that it will transmit. The values for the forward and backward maximum CPCS-SDU size indicated in the CONNECT message are less than or equal to the values specified by the calling user in the SETUP message.

If the called user does not include the CPCS-SDU size in the CONNECT message, the calling user assumes that the called user accepted the values of the forward and backward maximum CPCS-SDU size indicated in the SETUP message.

If the calling party cannot use the forward or backward maximum CPCS-SDU size indicated in the CONNECT message by the called party (for example, because the value negotiated by the called party is unacceptably small), the call is cleared with cause "AAL parameters can not be supported."

If the calling end station receives an ATM adaptation layer parameters IE in the CONNECT message that contains (1) octet groups other than the forward and backward maximum CPCS-SDU size and/or MID range, (2) a maximum SDU size that is greater than the maximum SDU size that was sent in

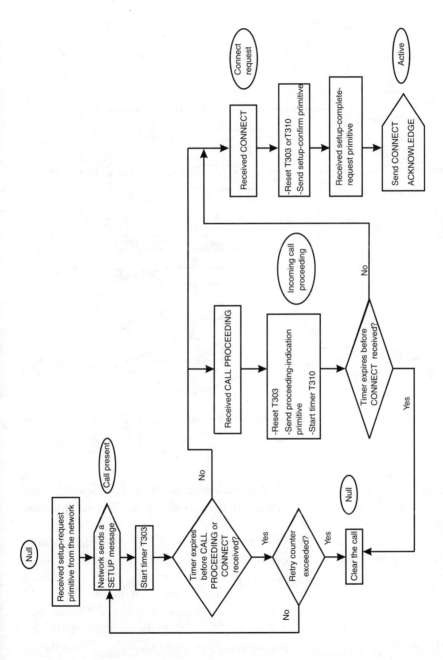

Figure 2.29 Connection establishment at the destination interface—network side without ALERTING message.

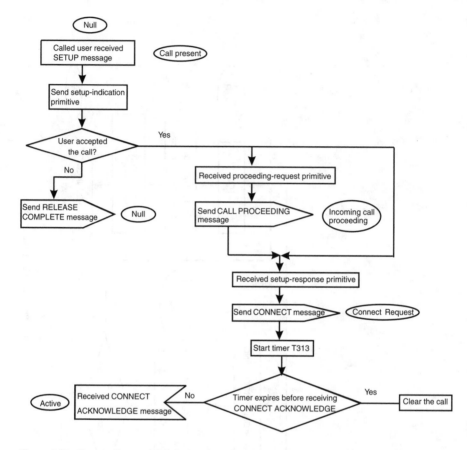

Figure 2.30 Connection establishment at the destination interface—user side without ALERTING message.

the SETUP message, or the forward or backward maximum CPCS-SDU size is missing, then the calling endpoint clears the call with cause "invalid information element contents."

The values of forward and backward maximum CPCS-SDU size resulting from AAL parameters negotiation are used by the AAL entities in the user plane. The AAL entity in the calling user equipment shall not send a CPCS-SDU size larger than the indicated value specified in the forward maximum CPCS-SDU size parameter and may allocate its internal resources based on the value indicated in the backward maximum CPCS-SDU size parameter. Similarly, the AAL entity in the called user equipment shall not send a CPCS-SDU size larger than the indicated value specified in the backward maximum CPCS-SDU size

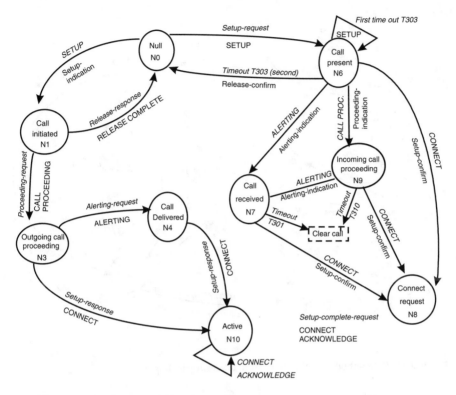

Figure 2.31 Connection-establishment state diagram—network side.

parameter and may allocate its internal resources based on the value indicated in the forward maximum CPCS-SDU size parameter.

2.4.4.2 MID Range Negotiation

When the called user receives the ATM adaptation layer parameters IE in the SETUP message that indicates AAL type 3/4, the called user checks the MID range value. If the called user cannot support the indicated MID range, but it can support a smaller range, it includes this IE in the CONNECT message containing the MID range that it can support.

The calling user either accepts the MID range contained in the CONNECT message or clears the call with cause "AAL parameters can not be supported." If the called user does not include the MID range in the CONNECT message, the calling user assumes that the called user accepts the MID range indicated by the calling user in the SETUP message.

If the calling endpoint receives an ATM adaptation layer parameters IE in the CONNECT message that contains octet groups other than the forward and

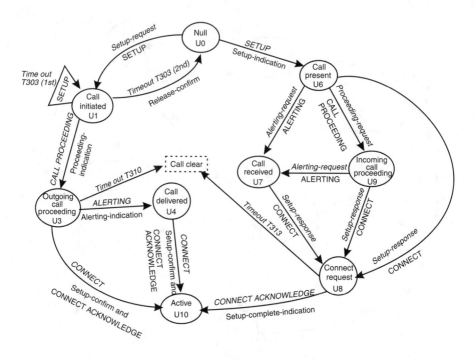

Figure 2.32 Connection-establishment state diagram—user side.

backward maximum CPCS-SDU size and/or MID range, or contains a MID range that is greater than the MID range that was sent in the SETUP message, it clears the call with cause "invalid information element contents."

2.4.5 Handling of the End-to-End Transit Delay Information Element

The end-to-end transit delay IE processing is a mandatory function in the network, whereas it is optional for the user. The purpose of the end-to-end transit delay IE is to indicate the maximum end-to-end transit delay acceptable for a call and to indicate the cumulative transit delay that can be expected for a connection.

The calling user may indicate a maximum end-to-end transit delay value to specify the end-to-end transit delay is acceptable for its connection. The transit delay across the link from the end station to the originating switch may be indicated by the calling user.

The network includes an end-to-end transit delay IE in the SETUP message sent to the called user if the calling user included this IE in the SETUP message. The called user may update the cumulative transit delay value received from the network. This is particularly important if the transmission line between the network boundary and the called terminal equipment causes substantial further delay. If the cumulative transit delay value exceeds the specified maximum end-to-end transit delay value, the called user may reject the call establishment request.

If the called user accepts the call, the end-to-end transit delay IE may be included in its CONNECT message specifying the final cumulative transit delay value for the call.

At the originating UNI, the inclusion of the end-to-end transit delay IE in the SETUP message by the calling user is optional. If it is included, both the cumulative transit delay subfield and the maximum end-to-end transit delay subfield need to be present. If this IE is received with one of the two fields only, the network handles the IE as a nonmandatory IE with content error.

At the destination UNI, the network includes the end-to-end transit delay IE if the calling user included this IE in the SETUP message. Both the cumulative transit delay subfield and the maximum end-to-end transit delay subfield need to be present.

When the called user receives this IE, it is recommended that it update the cumulative transit delay value received from the network. If the cumulative transit delay value exceeds the maximum end-to-end transit delay value specified by the calling user, it is also recommended that the called user reject the call with cause "quality of service not available."

If the SETUP message sent to the called user included an end-to-end transit delay IE, the called user may include an end-to-end transit delay IE in the CONNECT message specifying the final cumulative transit delay value for the call. The maximum end-to-end transit delay subfield is not included. If the network receives an end-to-end transit delay IE with this subfield, the field is discarded. The network does not check the correctness of the cumulative transit delay value provided.

The network includes an end-to-end transit delay IE in the CONNECT message sent to the calling user if the called user included this information in the CONNECT message. The maximum end-to-end transit delay subfield is not included.

2.5 Call/Connection Clearing

2.5.1 Call-Clearing Messages

The call-clearing messages used by Q.2931 are listed in Table 2.11.

Table 2.11
Call-Clearing Messages

Message Type	Significance	Direction	Description
RELEASE	Global	Both	Indicates that the connection has cleared and the sender is waiting to release the call reference
RELEASE COMPLETE	Local	Both	Indicates that the connection is internally cleared and the call reference is released

2.5.1.1 RELEASE

The RELEASE message is sent by the user to request the network to clear the end-to-end connection (if any) or is sent by the network to indicate that the end-to-end connection is cleared and that the receiving equipment can release the connection identifier and prepare to release its call reference value after sending RELEASE COMPLETE (see Figure 2.33).

2.5.1.2 RELEASE COMPLETE

The RELEASE COMPLETE message is sent by the user or the network to indicate that the end station sending the message has released its call reference value and, if appropriate, the connection identifier. The connection identifier, if released, is available for reuse. The end station releases its call reference value upon receiving the message. This message has local significance; however, it may carry information of global significance when used as the first call-clearing message (see Figure 2.34).

The cause IE is mandatory in the first call-clearing message, including the case when the RELEASE COMPLETE message is sent as a result of an error-handling condition.

Information Element	Type	Length
Cause	M	6–34

Figure 2.33 Information elements used in RELEASE message.

Information Element	Type	Length
Cause	O	4–34

Figure 2.34   Information elements used in RELEASE COMPLETE message.

2.5.2   Call-Clearing Procedures

Under normal conditions, call clearing is initiated when the user or the network sends a RELEASE message and follows the clearing procedures defined in this section. The only exception to this rule is when the user or network rejects a call/connection, for example, due to unavailability of a virtual channel by responding to a SETUP message with a RELEASE COMPLETE message.

The user initiates call clearing by sending a RELEASE message. Upon sending this message, it starts timer T308, disconnects the virtual channel, and enters the *release request* state. A VC is disconnected when the VC is no longer part of a B-ISDN virtual connection but is not yet available for use in a new virtual connection.

When user initiates normal call/connection clearing, the cause "normal clearing" is used in the first clearing message. The network enters the release request state upon receiving a RELEASE message. This message prompts the network to disconnect the virtual channel and to initiate procedures for clearing the network connection to the remote user. Once the virtual channel used for the call has been disconnected, the network sends a RELEASE COMPLETE message to the user, releases both the call reference and virtual channel, and enters the null state. A VC is released when the VC is not part of a B-ISDN virtual connection and is available for use in a new virtual connection. Similarly, a call reference is released and becomes available for reuse. The RELEASE COMPLETE message has only local significance and does not imply an acknowledgment of clearing from the remote user.

Upon receiving a RELEASE COMPLETE message, the user stops timer T308, releases the virtual channel, releases the call reference, and returns to the null state. If timer T308 expires for the first time, the user retransmits a RELEASE message to the network with the cause originally contained in the first RELEASE message, restarts timer T308, and remains in the release request state. In addition, the user may indicate a second cause IE with cause "recovery on timer expiry." If a RELEASE COMPLETE message is not received from the network before timer T308 expires the second time, the user places the virtual channel in a maintenance condition, releases the call reference, and returns to the null state (see Figure 2.35).

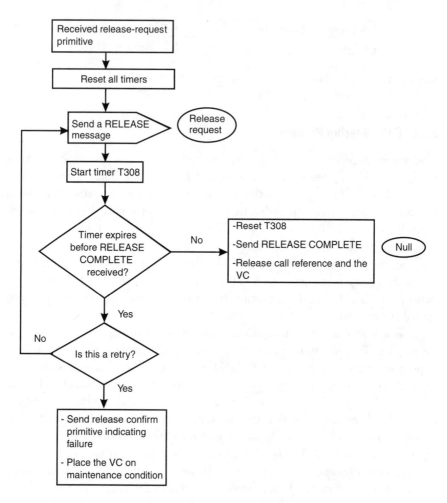

Figure 2.35 Call-clearing procedures.

Call clearing may also be initiated by the network. The network initiates call clearing by sending a RELEASE message, starting timer T308, disconnecting the virtual channel, and entering the *release indication* state. The user enters the release indication state upon receiving a RELEASE message. Once the virtual channel used for the call has been disconnected, the user sends a RELEASE COMPLETE message to the network, releases both its call reference and the virtual channel, and returns to the null state.

Upon receiving a RELEASE COMPLETE message, the network stops timer T308, releases both the virtual channel and call reference, and returns to the null state. If timer T308 expires for the first time, the network retransmits the RELEASE message to the user with the cause number originally contained in the first RELEASE message, starts timer T308, and remains in the release indication state. In addition, the network may indicate a second cause IE with cause "recovery on timer expiry." If a RELEASE COMPLETE message is not received from the user before timer T308 expires the second time, the network places the virtual channel in a maintenance condition, releases the call reference, and returns to the null state.

2.5.3 Call/Connection Clearing—State Diagrams

State transitions that take place at the user and the network sides for call/connection clearing are illustrated in Figures 2.36 and 2.37, respectively.

2.5.4 Clear Collision

When both sides simultaneously transfer RELEASE messages related to the same call reference value, clear collision can occur. If the user receives a RELEASE message while in the release request state, it stops timer T308, releases the call reference and the virtual channel, and enters the null state (without sending or receiving a RELEASE COMPLETE message). Similarly, if the network receives a RELEASE message while in the release indication state, the network stops timer T308, releases the call reference and virtual channel, and enters the null state (see Figure 2.38).

2.6 Restart Procedure

The restart procedure is used to return a virtual channel, all virtual channels in a virtual path, or all virtual channels controlled by the signaling virtual channel to the idle condition. The procedure is usually invoked when the other side of the interface does not respond to other call control messages or a failure has occurred (e.g., following the expiry of timer T308 due to the absence of response to a clearing message). It may also be initiated as a result of local failure, maintenance action, or misoperation.

2.6.1 Messages and Information Elements

Restart procedures use the following two messages with a global call reference.

2.6.1.1 RESTART

The RESTART message is sent by the user or the network to request the recipient to restart (i.e., return to an idle condition) the indicated virtual channel, all

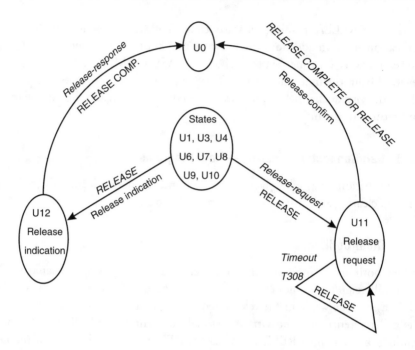

Figure 2.36 Connection-clearing state diagram—user side.

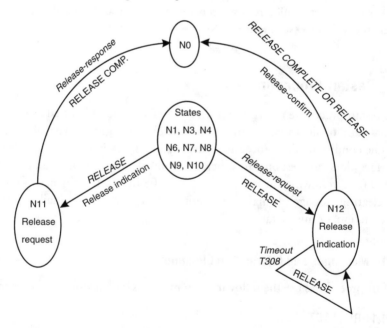

Figure 2.37 Connection-clearing state diagram—network side.

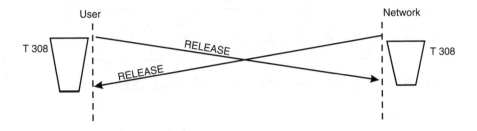

Figure 2.38 Clear collision.

virtual channels in the indicated virtual path connection, or all virtual channels controlled by the signaling virtual channel (see Figure 2.39).

A connection identifier IE is included, when necessary, to indicate the particular virtual channel(s) to be restarted. The restart indicator IE is described in Section 2.6.1.3.

2.6.1.2 RESTART ACKNOWLEDGE

The RESTART ACKNOWLEDGE message is sent to acknowledge the receipt of a RESTART message and to indicate that the requested restart is complete (see Figure 2.40).

A connection identifier IE is included when necessary to indicate the particular virtual channel(s) that have been restarted.

2.6.1.3 Restart Indicator Information Element

The restart indicator IE is used to identify the class of the facility to be restarted. Its format is illustrated in Figure 2.41.

The class field allows the specification of VCs to be restarted:

1. *Indicated virtual channel*—This requires the connection identifier IE to be included to indicate the VC to be restarted.
2. *Indicated switched virtual path or all virtual channels in the indicated VPC that are controlled via the signaling virtual channel on which the*

Information Element	Type	Length
Connection identifier	O	4–9
Restart indicator	M	5

Figure 2.39 Information elements used in RESTART message.

Information Element	Type	Length
Connection identifier	O	4–9
Restart indicator	M	5

Figure 2.40 Information elements used in RESTART ACKNOWLEDGE message.

Field						Octets
1	0	0	0	0	Class	5
ext.			Spare			

Figure 2.41 Restart indicator IE.

RESTART message is sent—This requires the connection identifier IE to be included to indicate the VPI in which all virtual channels are to be restarted.

3. *All virtual channels controlled by the layer 3 entity that sends the RESTART message*—This causes all virtual channels controlled by the point-to-point signaling channel are to be restarted.

The call reference flag of the global call reference applies to restart procedures. In the case when both sides of the interface initiate simultaneous restart requests, they are handled independently. In the case when the same user plane virtual channel(s) are specified, they are not considered free for reuse until all the relevant restart procedures are completed. In the RESTART message, the call reference flag of the global call reference is set to 0. In the RESTART ACKNOWLEDGE message sent in response to a RESTART message, the call reference flag of the global call reference is set to 1.

2.6.2 Sending a RESTART Message

A RESTART message is sent by the network or user equipment to return virtual paths or virtual channels to the idle condition. The restart indicator IE is present in the RESTART message to indicate whether an indicated virtual channel, virtual path, or all user plane virtual channels in the indicated VPC controlled via the signaling virtual channel in which the RESTART message is sent, or all virtual channels controlled by the layer 3 entity are to be restarted. If the restart indicator IE is coded as indicated virtual channel or all user plane virtual channels in the indicated VPC controlled via signaling virtual channel in which the RESTART message is sent, the connection identifier IE needs to be present to indicate which virtual channel or virtual path is to be returned to the idle condition. If the restart indicator IE is coded as all virtual channels controlled by the layer 3 entity that sends the RESTART message, then the connection identifier IE is not included.

Upon transmitting a RESTART message, the sender enters the *restart request* state, starts timer T316, and waits for a RESTART ACKNOWLEDGE message. No other RESTART messages can be sent until a RESTART ACKNOWLEDGE is received or timer T316 expires. Receiving a RESTART ACKNOWLEDGE message causes the stopping of timer T316 and indicates that the virtual channel(s) and associated resources including the call reference value(s) can be reused. After releasing the virtual channel and call reference value, the recipient enters the null state.

The RESTART and RESTART ACKNOWLEDGE message contains the global call reference value (all 0's) to which the restart request state is associated. Calls associated with the restarted virtual channels are cleared towards the remote parties using cause "temporary failure."

If a RESTART ACKNOWLEDGE message is not received prior to the expiry of timer T316, one or more subsequent RESTART messages may be sent until a RESTART ACKNOWLEDGE message is returned. While timer T316 is running, the virtual channel(s) being restarted cannot be used to support new calls requested using the call setup procedures. The number of consecutive unsuccessful restart attempts has a default limit of two. When this limit is reached, the originator of the RESTART message makes no further restart attempts and enters the null state. An indication is provided to the appropriate maintenance entity. The virtual channel(s) is considered to be in an out-of-service condition until maintenance action has been taken. If a RESTART ACKNOWLEDGE message is received, indicating a different set of virtual channels from the set indicated in the RESTART message, the RESTART ACKNOWLEDGE message is discarded.

Figure 2.42 illustrates the sequence of events that takes place when a RESTART message is sent.

2.6.3   Receiving a RESTART Message

Upon receiving a RESTART message, the recipient enters the *restart* state associated with the global call reference and starts timer T317. It then initiates the appropriate internal actions to return the specified virtual path or virtual channels to the idle condition and releases all call references associated with the specified virtual channels. Upon completion of internal clearing, timer T317 is stopped, a RESTART ACKNOWLEDGE message is transmitted to the originator, and the null state (REST 0) is entered. If timer T317 expires prior to

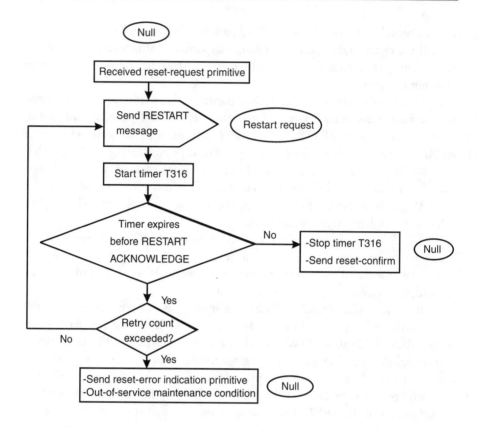

Fig 2.42 Sending RESTART message.

completion of internal clearing, an indication is sent to the maintenance entity (i.e., a primitive shall be transmitted to the system management entity) and the null state (REST 0) is entered. The RESTART ACKNOWLEDGE message indicates a restart indicator IE containing the same information as the received information in the related RESTART message. In addition, the RESTART ACKNOWLEDGE message contains a connection identifier IE containing the same information as received in the related RESTART message. Calls associated with restart user plane virtual channels are cleared towards the remote parties using cause "temporary failure."

Even if all the specified virtual channels are in the idle condition or already in the process of restarting to the idle condition, the receiving entity is required to transmit a RESTART ACKNOWLEDGE message to the originator upon receiving a RESTART message.

If the restart indicator IE is coded as all virtual channels controlled by the layer 3 entity which sends the RESTART message, then all calls on the interface(s) associated with the signaling virtual channel are cleared.

If semipermanent connections established by the management procedures are implicitly specified (by specifying all virtual channels that are controlled by the layer 3 entity that sends the RESTART message or all user plane virtual channels in the indicated VPC controlled via the signaling virtual channel in which the RESTART message is sent), no action is taken on these virtual channels, but a RESTART ACKNOWLEDGE message is returned containing the appropriate indications.

If semipermanent connections established by management procedures or reserved VPCI/VCIs (e.g., the point-to-point signaling virtual channel) are explicitly specified (by including a connection identifier IE in the RESTART message), no action is taken on these virtual channels and a STATUS message may as an option be returned with cause "identified channel does not exist," optionally indicating in the diagnostic field the virtual channel or virtual path that could not be handled. In summary, entities released as a result of the restart procedures are

- Virtual channels established by Q.2931 procedures;
- All resources associated with the released virtual channel (e.g., call reference value).

Entities that are not released as a result of the restart procedures are

- Permanent connections established by a network;
- Management system reserved virtual channels (e.g., point-to-point signaling virtual channel).

Figure 2.43 illustrates the sequence of events that takes place when responding to a RESTART message.

2.7 Handling of Error Conditions

All messages that use the protocol discriminator Q.2931 user-network call control message must pass the checks described in this section. Detailed error-handling procedures are implementation dependent and may vary from network to network.

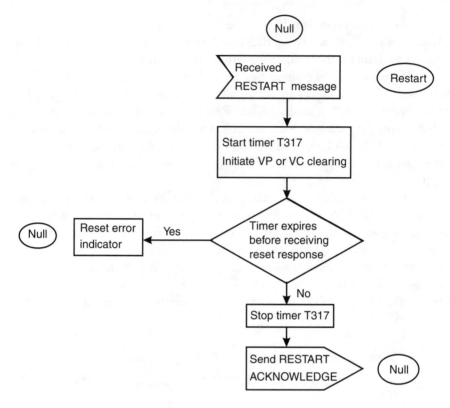

Figure 2.43 Receipt of RESTART message.

2.7.1 Protocol Discrimination Error

A message received with a *protocol discriminator* coded as other than a Q.2931 user-network call control message is ignored. In this context, ignore means doing nothing, as if the message had never been received.

2.7.2 Message Too Short

If a message is received short of its *message length* IE, it is ignored.

2.7.3 Call Reference Error

If the *call reference* IE octet 1, bits 5 through 8, is not equal to 0000, then the message is ignored. If the call reference IE octet 1, bits 1 through 4, indicates a length other than 3 octets, the message is ignored.

The call reference procedural errors are as follows:

1. Whenever a message other than SETUP, RELEASE COMPLETE, STATUS ENQUIRY, or STATUS is received with an unrecognized call reference (i.e., does not relate to an active call or to a call in progress), the receiver initiates call clearing by sending a RELEASE COMPLETE message.
2. When a RELEASE COMPLETE message is received with an unrecognized call reference, no action is taken.
3. When a SETUP message is received with an unrecognized call reference and the call reference flag is incorrectly set to 1, this message is ignored.
4. When any message except RESTART, RESTART ACKNOWLEDGE, or STATUS is received using the global call reference, no action is taken on this message.

2.7.4 Message Type or Message Sequence Errors

The error procedures in this section apply only if the flag in the message compatibility instruction indicator is set to "message instruction field not significant." If it is set to "follow explicit instructions," the procedures in Section 2.10 take precedence.

Whenever an unexpected message, except RELEASE, RELEASE COMPLETE, or an unrecognized message is received in any state other than the *null state*, no state change occurs and a STATUS message is returned with one of the following causes: "message type nonexistent or not implemented" or "message not compatible with call state."

2.7.5 General Information-Element Errors

If an IE is repeated in a message and repetition is not permitted, only the contents of the IE appearing first is handled and all subsequent repetitions of the IE are ignored. When repetition of IEs is permitted, and if the limit on repetition of IEs is exceeded, the contents of IEs appearing first up to the limit of repetitions are handled and all subsequent repetitions of the IE are ignored.

If the user or the network receives an IE with the coding standard that is not supported by the receiver, this IE is treated as an IE with a content error (cf. Section 2.7.7.2).

2.7.6 Mandatory Information-Element Errors

2.7.6.1 Mandatory Information Element Missing

When a message other than SETUP, RELEASE, or RELEASE COMPLETE is received with one or more mandatory IEs missing, no action is taken on the message and no state change occurs. A STATUS message is returned in this case with cause "mandatory information element is missing."

When a SETUP message is received with one or more mandatory IEs missing, a RELEASE COMPLETE message with cause "mandatory information element is missing" is returned.

When a RELEASE message is received with the cause IE missing, the actions taken are the same as if a RELEASE message with cause "normal, unspecified" was received, with the exception that the RELEASE COMPLETE message is sent on the local interface with cause "mandatory information element is missing."

When a RELEASE COMPLETE message is received with the cause IE missing, it is assumed that a RELEASE COMPLETE message was received with cause "normal, unspecified."

2.7.6.2 Mandatory Information Element Content Error

The error procedures in this section apply only if the flag in the instruction field is set to "IE instruction field not significant." If it is set to "follow explicit instructions," the procedures in Section 2.10 take precedence.

When a message other than SETUP, RELEASE, or RELEASE COMPLETE is received with one or more mandatory IEs with invalid content, no action is taken on the message and no state change occurs. A STATUS message is returned with cause "invalid information element contents."

When a SETUP message is received with one or more mandatory IEs with invalid content, a RELEASE COMPLETE message with cause "invalid information element contents" is returned.

When a RELEASE message is received with invalid content of the cause IE, the actions taken are the same as if a RELEASE message with cause "normal, unspecified" was received, with the exception that the RELEASE COMPLETE message is sent on the local interface with cause "invalid information element contents."

When a RELEASE COMPLETE message is received with invalid content of the cause IE, it is assumed that a RELEASE COMPLETE message was received with cause "normal, unspecified."

IEs with a length exceeding the maximum length are treated as IEs with content error.

2.7.7 Nonmandatory Information-Element Errors

The error procedures in this section apply only if the flag in the instruction field is set to "IE instruction field not significant." If it is set to "follow explicit instruction," the procedures in Section 2.10 take precedence.

2.7.7.1 Unrecognized Information Element

When a message is received with one or more unrecognized IEs, the receiving entity proceeds as follows. When the received message is other than RELEASE or RELEASE COMPLETE, a STATUS message may be returned containing one cause IE. The STATUS message indicates the call state of the receiver after taking action on the message. The cause IE contains the cause "information element nonexistent or not implemented," and the diagnostic field, if present, contains the IE identifier for each IE that was unrecognized. Subsequent actions are determined by the sender of the unrecognized IEs.

If a clearing message contains one or more unrecognized IEs, the error is reported to the local user as follows:

1. When a RELEASE message is received with one or more unrecognized IEs, a RELEASE COMPLETE message with the cause "information element nonexistent or not implemented" is returned. The cause IE diagnostic field, if present, contains the IE identifier for each IE that was unrecognized.

2. When a RELEASE COMPLETE message is received with one or more unrecognized IEs, no action is taken on the unrecognized information.

2.7.7.2 Nonmandatory Information Element Content Error

When a message is received with one or more nonmandatory IEs with invalid content, regular processing takes place on the message using the IEs that are recognized and have valid contents. A STATUS message may be returned indicating the call state of the receiver after taking the appropriate action on the message. The cause IE contains the cause "invalid information element contents," and the diagnostic field, if present, contains the IE identifier for each IE that has invalid contents.

IEs with a length exceeding the maximum length are treated as an IE with content error. For access IEs, cause "access information discarded" is used instead of cause "invalid information element contents."

2.7.7.3 Unexpected Recognized Information Element

When a message is received with a recognized IE that is not defined to be contained in that message, the receiving entity treats the IE as an unrecognized IE and follow the procedures defined in Section 2.7.7.1.

2.8 Signaling AAL Connection Reset

Whenever a Q.2931 entity is informed of a spontaneous signaling AAL reset by means of the AAL-ESTABLISH-indication primitive, the following procedures apply:

1. For calls in the clearing phase (states N11, N12, U11, and U12), no action is taken.
2. For calls in the establishment phase, the states N1, N3, N4, N6, N7, N8, N9, U1, U3, U4, U6, U7, U8, and U9 are maintained. Optionally, the status enquiry procedure may be invoked.
3. Calls in the active state are maintained and the entity invokes the status enquiry procedures.

Whenever a Q.2931 entity is notified of its signaling AAL connection release by means of the AAL-RELEASE-indication primitive, the following procedure applies:

1. Any calls that are not in the active state are cleared locally.
2. If there is at least one call in the active state controlled by the released signaling AAL connection and timer T309 is not running, then timer T309 is started.

The Q.2931 entity requests signaling AAL re-establishment by sending an AAL-ESTABLISH-request primitive.

When informed of signaling AAL re-establishment by means of the AAL-ESTABLISH-confirm primitive, timer T309 is stopped and the status enquiry procedure is performed to verify the call state of the peer entity per each call/connection.

If timer T309 expires prior to signaling AAL re-establishment, the network clears the network connection and the call to the remote user with cause "destination out of order," disconnects and releases the bearer virtual channel, releases the call reference, and enters the null state. If timer T309 expires prior to signaling AAL re-establishment, the user disconnects and releases the virtual channel, releases the call reference, and enters the null state. The user may clear the attached internal connection (if any) with cause "destination out of order."

2.9 Status Enquiry

A STATUS ENQUIRY message may be sent to check the correctness of a call state at a peer entity. This may, in particular, apply to procedural error conditions described in Section 2.7. In addition, whenever an indication is received from the signaling AAL that a disruption has occurred at the data-link layer, a STATUS ENQUIRY message is sent to check the correctness of the call state at the peer entity.

2.9.1 Messages and Information Elements

2.9.1.1 STATUS

The STATUS message is sent by the user or the network in response to a STATUS ENQUIRY message or at any time to report certain error conditions. If this message is sent with the global call reference, the *global interface* state is indicated by the *call state* IE (see Figure 2.44).

The STATUS message uses the global call reference if it is sent in response to a message with the global call reference. The call reference flag in this STATUS message is coded to 0 if the call reference flag in the received message was set to 1, and it is coded to 1 if the call reference flag in the received message was set to 0.

2.9.1.2 STATUS ENQUIRY

The STATUS ENQUIRY message is sent by the user or the network at any time to solicit a STATUS message from the peer layer 3 entity. Sending a STATUS message in response to a STATUS ENQUIRY message is mandatory.

Information Element	Type	Length
Call state	M	5
Cause	M	6–34

Figure 2.44 Information elements used in STATUS message.

2.9.1.3 Call State Information Element

The call state IE is used to describe the current status of a call/connection. Its format is illustrated in Figure 2.45.

The call states in Q.2931 are classified as user state or network state and listed as shown in Table 2.12.

Table 2.12
Q.2931 Call States

User State		Network State	
U0	Null	N0	Null
U1	Call initiated	N1	Call initiated
U3	Outgoing call proceeding	N3	Outgoing call proceeding
U4	Call delivered	N4	Call delivered
U6	Call present	N6	Call present
U7	Call received	N7	Call received
U8	Connect request	N8	Connect request
U9	Incoming call proceeding	N9	Incoming call proceeding
U10	Active	N10	Active
U11	Release request	N11	Release request
U12	Release indication	N12	Release indication

In addition, there are the global interface state values shown in Table 2.13.

2.9.2 Status Enquiry Procedure

Upon sending the STATUS ENQUIRY message, timer T322 is started in anticipation of receiving a STATUS message. While timer T322 is running, only one outstanding request for the call state information exists. Therefore, if timer

Field		Octet
0 0	Call state value/global interface state value	5
Spare		

Figure 2.45 Call state IE format.

Table 2.13

Q.2931 Global Interface States

State	Description
REST 0	Null
REST 1	Restart request
REST 2	Restart

T322 is already running, it is not restarted. If a clearing message is received before timer T322 expires, timer T322 is stopped and call clearing continues.

Upon receipt of a STATUS ENQUIRY message, the receiver responds with a STATUS message reporting the current call state (of an active call, a call in progress, or the null state if the call reference does not relate to an active call or to a call in progress) and the cause "response to STATUS ENQUIRY." Receiving a STATUS ENQUIRY message does not cause a state change.

If timer T322 expires and no STATUS message was received, the STATUS ENQUIRY message may be retransmitted one or more times until a response is received. The number of times the STATUS ENQUIRY message is retransmitted is an implementation-dependent value. If (following the maximum number of retransmissions of the STATUS ENQUIRY message) a STATUS message is not received before T322 expires for the last time, the call is cleared at the local interface with cause "temporary failure." If appropriate, the network also clears the network connection using cause "temporary failure."

Figure 2.46 illustrates the sequence of events that takes place when status enquiry procedures are invoked.

2.9.3 Receiving a STATUS Message

Upon receiving a STATUS message reporting an incompatible state, the receiving entity does the following:

1. Clears the call by sending the appropriate clearing message with cause "message not compatible with call state;"
2. Takes other implementation dependent actions that attempt to recover from a mismatch.

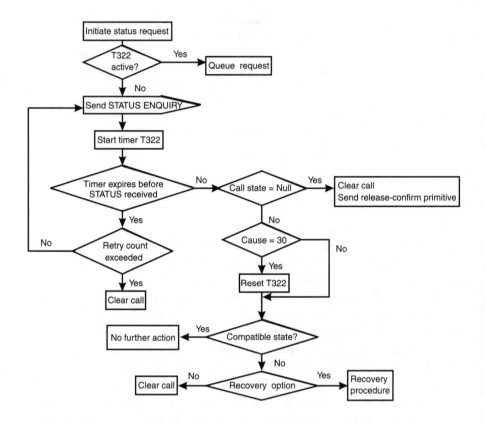

Figure 2.46 Status enquiry procedures.

Except for the following rules, the determination of which states are incompatible is left as an implementation decision:

1. If a STATUS message indicating any call state except the null state is received in the null state, the receiving entity sends a RELEASE COMPLETE message with cause "message not compatible with call state" and remains in the null state.

2. If a STATUS message indicating any call state except the null state is received in the release request or release indication state, no action is taken.

3. If a STATUS message indicating the null state is received in any state except the null state, the receiver releases all resources and moves into the null state.

When in the null state, the receiver of a STATUS message indicating the null state takes no action other than to discard the message and remains in the null state.

A STATUS message may be received indicating a compatible call state but containing one of the following causes:

- "mandatory information element is missing;"
- "message type nonexisting or not implemented;"
- "information element nonexistent or not implemented;"
- "invalid information element contents;"
- "message not compatible with call state."

In these cases, the actions taken are an implementation option. If other procedures are not defined, the receiver clears the call using the cause specified in the received STATUS message (see Figure 2.47).

2.10 Error Procedures With Explicit Action Indication

The procedures described in this section are used if the flag of the message compatibility instruction indicator or IE instruction field is set to "follow explicit instructions."

2.10.1 Unexpected or Unrecognized Message Type

The following procedures are used when an unexpected or unrecognized message type is received in any state other than the null state.

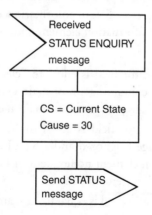

Figure 2.47 Receiving a STATUS message.

If the action indicator bits of the instruction field of a message type IE are set to "clear call" in any state other than the release request and release indication state, the call is cleared with cause "message type nonexistent or not implemented" or cause "message not compatible with call state."

If the action indicator bits of the instruction field of a message type IE are set to "discard and ignore," the message is ignored. If these bits are set to "discard, and report status," no action is taken on the message, but a STATUS message is sent with cause "message type nonexistent or not implemented" or "message not compatible with call state." If the message action indicator bits are set to an *"undefined (reserved) value,"* the receiver handles the message as if the message action indicator bits had been set to "discard, and report status."

2.10.2   Information-Element Errors

When a message other than a RELEASE or RELEASE COMPLETE message is received with one or more unexpected IEs, unrecognized IEs, or IEs with unrecognized contents, the receiving entity examines the IE action indicator and follows the procedures described next, as appropriate.

When a RELEASE message is received with one or more IEs in error, a RELEASE COMPLETE message with cause "information element nonexistent or not implemented" or with cause "invalid information element contents" is returned. When a RELEASE COMPLETE message is received with one or more IEs in error, no action is taken on the IEs in error. The message is processed as if received without the IE in error.

If more than one IE is received in error, only one response is given. The response depends on the action indicator field according to the following order of priority: "clear call" (highest priority), "discard message and report status," "discard message and ignore," "discard information element, proceed and report status," and "discard information element and proceed."

1. If the action indicator field is equal to "clear call," the call is cleared with the cause IE containing "information element nonexistent or not implemented" or cause "invalid information element contents."

2. If the action indicator field is equal to "discard message and report status," the message is ignored and a STATUS message is sent with a cause "information element nonexistent or not implemented" or cause "invalid information element contents."

3. If the action indicator field is equal to "discard message," the message is ignored.

4. If the action indicator field is "discard information element, proceed and report status," and if the message contains sufficient information to proceed, the following applies: If the action indicator field specifies "discard information element, proceed and report status," the IE is discarded, the rest of the message is processed and a STATUS message is returned indicating the call state of the receiver after taking action on the message with cause "information element nonexistent or not implemented" or cause "invalid information element contents."

5. If the action indicator field is equal to "discard information element and proceed," the IE is ignored and the message is processed as if the IE was not received. No STATUS message is sent.

6. If the action indicator field is equal to "undefined (reserved) value," the receiver handles the IE as if the action indicator field had been set to "discard information element, proceed, and report status."

2.10.3 Handling of Messages With Insufficient Information

If there is not sufficient information to handle a message, the corresponding procedures defined previously apply if the flag of the message compatibility instruction indicator is set to "message instruction field not significant" or "follow explicit instruction" with a cause IE with cause "mandatory information element is missing."

2.11 List of Timers

The description of the timers in Tables 2.14 and 2.15 offers a brief summary. The precise details are found in the earlier subsections.

2.12 Examples

In this section, three examples are presented on the use of signaling messages and information elements for three different types of services.

2.12.1 LAN Emulation Over ATM

Ethernet and Token Ring LAN stations use connectionless service. The ATM Forum LAN Emulation service is designed to provide the appearance of such a connectionless service to the participating end stations in an ATM network.

Table 2.14
Timers in the Network Side

Timer	Default Value	Call State	Cause for Start	Cause for Normal Stop	Action at the First Expiry	Action at the Second Expiry	Implementation
T301	Min. 3 min.	Call received	ALERT received	CONNECT received	Clear call	Not applicable	Mandatory (Note 1)
T303	4 sec.	Call present	SETUP sent	ALERT, CONNECT, RELEASE COMPLETE, or CALL PROCEEDING received	Resend SETUP, restart T303	Clear network connection; enter null state	Mandatory
T308	30 sec.	Release indication	RELEASE sent	RELEASE COMPLETE or RELEASE received	Resend RELEASE; restart T308	Place bearer virtual channel in maintenance condition. Release call reference and enter null state (Note 2)	Mandatory
T309	10 sec.	Any stable state	SAAL disconnection	SAAL reconnected	Clear network connection; release connection and call reference	Not applicable	Mandatory
T310	30 sec.	Incoming call proceeding	CALL PROCEEDING received	ALERT, CONNECT, or RELEASE received	Clear call	Not applicable	Mandatory

Table 2.14 (Continued)

Timer	Default Value	Call State	Cause for Start	Cause for Normal Stop	Action at the First Expiry	Action at the Second Expiry	Implementation
T316	2 min.	Restart request	RESTART sent	RESTART ACKNOWLEDGE received	RESTART may be sent several times	RESTART may be sent several times	Mandatory for point-to-point access configuration
T317	(Note 2)	Restart	RESTART received	Internal clearing of call references	Maintenance notification	Not applicable	Mandatory for point-to-point access configuration
T322	4 sec.	Any call state	STATUS ENQUIRY sent	STATUS, RELEASE, or RELEASE COMPLETE received	STATUS ENQUIRY may be re-sent several times	STATUS ENQUIRY may be re-sent several times	Mandatory

(1) The network may already have applied an internal alerting supervision timing function; e.g., incorporated within call control. If such a function is known to be operating on the call, then timer T301 is not used.

(2) The value of timer T317 is implementation dependent, but a value that is less than the likely values of T316 in peer implementation shall be chosen.

Table 2.15
Timers in the User Side

Timer	Default Value	Call State	Cause for Start	Cause for Normal Stop	Action at the First Expiry	Action at the Second Expiry	Implementation
T301	Min. 3 min.	Call delivered	ALERT received	CONNECT received	Clear call	Not applicable	Mandatory if symmetrical procedures are supported
T303	4 sec.	Call initiated	SETUP sent	ALERT, CONNECT, RELEASE COMPLETE or CALL PROCEEDING received	Resend SETUP; restart T303	Clear internal connection; enter null state	Mandatory
T308	30 sec.	Release request	RELEASE sent	RELEASE COMPLETE or RELEASE received	Resend RELEASE; restart T308	Place bearer virtual channel in maintenance condition. Release call reference and enter null state	Mandatory
T309	10 sec.	Any stable state	SAAL disconnection	SAAL reconnected	Clear network connection; release connection and call reference	Not applicable	Mandatory

Table 2.15 (Continued)

Timer	Default Value	Call State	Cause for Start	Cause for Normal Stop	Action at the First Expiry	Action at the Second Expiry	Implementation
T310	30 to 120 sec.	Outgoing call proceeding	CALL PROCEEDING received	ALERT, CONNECT, or RELEASE received	RELEASE sent	Not applicable	Mandatory
T313	4 sec.	Connect request	CONNECT sent	CONNECT ACKNOWLEDGE received	RELEASE sent	Not applicable	Mandatory
T316	2 min.	Restart request	RESTART sent	RESTART, ACKNOWLEDGE received	RESTART may be sent several times	RESTART may be sent several times	Mandatory for point-to-point access configuration
T317	Implementation dependent	Restart	RESTART received	Internal clearing of call references	Maintenance notification	Not applicable	Mandatory for point-to-point access configuration
T322	4 sec.	Any call state	STATUS ENQUIRY sent	STATUS, RELEASE, or RELEASE COMPLETE received	STATUS ENQUIRY may be re-sent several times	STATUS ENQUIRY may be re-sent several times	Mandatory

The LAN Emulation service enables end stations to connect to an ATM network while the software applications interact as if they are attached to a traditional LAN.

In a switched virtual connection (SVC) environment, the LAN Emulation entities set up connections between each other using UNI signaling. These connections use best-effort QoS. There are six types of connections defined in the LAN Emulation service. Each type of connection is distinguished by the B-LLI code values and ATM addresses, and a separate virtual channel connection (VCC) is established between different LAN Emulation entities.

The values of various information elements used to characterize and ATM Forum LAN Emulation service are defined in Table 2.16.

2.12.2   Video on Demand

Video on demand (VoD) is a video service in which the end user has a predetermined level of control on selection of the material viewed and the time of viewing. Video connections are established on demand via user-network signaling. ATM Forum's VoD specification uses MPEG-2 (Motion Pictures Expert Group) bit streams over ATM to support this service. In particular, the specification specifies the following:

- AAL requirements;
- The encapsulation of MPEG-2 transport streams into AAL-5 PDUs;
- The ATM signaling and ATM connection control requirements;
- The traffic characteristics;
- The quality of service characteristics.

The ITU H.310 specifications define how ATM virtual circuits are established and use a control protocol H.245 for capability negotiation and mode setting whereby the actual means of packetizing the MPEG-2 transport packets into ATM AAL-5 protocol data units is specified in H.222.1. The ATM Forum VoD specification (hereafter referred to as AMS) is similar to the ITU framework.

The H.310 requires a two-phase call setup sequence. Initially, the control circuit is set up between peer audiovisual terminals. After these terminals exchange their capabilities through the control protocol, the sending side initiates the ATM VC setup procedure. This SETUP message includes a corrector carried in the *generic transport identifier* IE (see Figure 2.48).

The ATM virtual connection for VoD uses CBR ATM traffic type with AAL-5. The generic identifier transport signaling capability allows the generation and transport of information used by VOD applications across the ATM

Table 2.16
Q.2931 Signaling IEs Used in ATM Forum LAN Emulation Service

Information Element	Service Type = UBR	Service Type = ABR	Service Type = nrtVBR
AAL parameters	AAL Type = AAL-5 Forward CPCS SDU size - Control: 1516; Data: based on SDU size Backward CPCS SDU size SSCS type = Null		
ATM traffic descriptor	PCR line rate Best-effort indicator Frame discard indicator	PCR = desired value MCR = desired value Frame discard indicator	PCR = desired value SCR and MBS = desired value Frame discard indicator
Minimum acceptable ATM traffic descriptor	PCR = desired value		
ABR setup parameters		ABR setup parameter values	
Broadband bearer capability	Bearer class = BCOB-X Susceptibility to clipping = 0 User plane connection = point-to-point (PtP) or point-to-multipoint (PtMp)	Bearer class = BCOB-X Broadband transfer capability = $0 \times 0C$ Susceptibility to clipping = 0 User plane connection = PtP	Bearer class = BCOB-X Susceptibility to clipping = 0 User plane connection = PtP or PtMp

Table 2.16 (Continued)

Information Element	Service Type = UBR	Service Type = ABR	Service Type = nrtVBR

Broadband low-layer information	For nonmultiplexed VCCs
	User layer 3 = TR 9577 - SNAP identifier
	SNAP id = five octets (three octets ATM Forum OUI)
	PID - 0001 for direct VCC; control direct VCC and control distribute
	    0002 for Ethernet/IEEE 802.3 LE data direct VCC
	    0003 for 802.5 LE data direct VCC
	    0004 for 802.3 multicast send and multicast forward
	    0005 for 802.5 multicast send and multicast forward
	
	For LLC-multiplexed VCCs
	User information layer 2 = LAN LLC		

QoS parameter	QoS class forward: 0 QoS class backward: 0	QoS class forward: 0 QoS class backward: 0	QoS class forward: 0 or network-specific QoS class backward: 0 or network-specific
Calling and called party number	ATM endsystem address Type of number = unknown Addressing/numbering plan = ATM endsystem address		

Field	Octets
Identifier related standard/application	5
Identifier type	6*
Identifier length	6.1
Identifier value	6.2–6.n

Figure 2.48 Generic transport identifier IE.

networks, mainly by providing session correlation and/or resource identifiers. The generic transport identifier IE may be repeated in a message. The format of the generic transport IE is shown in Figure 2.48.

The two identifiers that are at currently specified are session and resource identifiers.

Table 2.17 summarizes the IE contents values used in the VoD service.

2.12.3 IP Over ATM

The classical IP over ATM service is described in RFC 1577. Similar to ATM Forum's LAN Emulation service, classical IP over ATM provides the basic framework to emulate a LAN segment connecting IP end stations and IP links that interconnect routers over ATM. Unlike LAN Emulation service, classical IP over ATM service can be used within an IP network only. The ATM network emulates a LAN segment for the following:

1. LANs that run IP;
2. Local-area backbones between existing (non-ATM) LANs that run IP;
3. Dedicated circuits or frame relay PVCs between IP routers.

The classical model here refers to the treatment of the ATM host adapter as a network interface to the IP protocol stack operating in a LAN-based paradigm. Characteristics of the classical model are as follows:

- LLC/SNAP encapsulation (RFC 1483) is the default packet format for IP datagrams;
- End-to-end IP routing architecture is the same;
- IP addresses are resolved to ATM addresses using an ATMARP service within a logical IP subnet;
- Each virtual connection connects two IP members.

In a switched virtual connection environment, ATM virtual channel connections are dynamically established and released as needed. ATM call control

Table 2.17
Q.2931 Signaling Information Elements Specific To H.310 or ATM Forum VoD

Information Element	H.310 Control VC	H.310 RAST-1 A/V VC	H.310 ROT/SOT-1 A/V VC	AMS VoD
AAL parameters	AAL type = AAL-5 Forward CPCS SDU size Backward CPCS SDU size SSCS type = frame relay	AAL type = AAL-5 Forward CPCS SDU size Backward CPCS SDU size SSCS type = Null		
ATM traffic descriptor	Forward PCR and backward PCR = 167 cells per second, which corresponds to 64 Kbps Sustainable cell rate = implementation-specific Maximum burst size = 2,048	Forward PCR = implementation-specific Backward PCR = implementation-specific Note: Forward and backwards are set according to which terminal sends the original SETUP message		
Broadband bearer capability	Bearer Class = BCOB-X C Broadband transfer capability = BTC10 User plane connection = PtP	Bearer class = BCOB-X Broadband transfer capability = CBR User plane connection = PtP		
Broadband low-layer information	User layer 3 = H.310 Terminal type = ROT/SOT/RAST Forward multiplexing = NO Backward multiplexing = NO	User layer 3 = H.310 Terminal type = RAST Forward multiplexing = TS Backward multiplexing = TS	User layer 3 = H.310 Terminal type = ROT/SOT Forward multiplexing = TS Backward multiplexing = Null	

Table 2.17 (Continued)

Information Element	H.310 Control VC	H.310 RAST-1 A/V VC	H.310 ROT/SOT-1 A/V VC	AMS VoD
Broadband high-layer information				High layer = Vendor-specific OUI type = ATM Forum Application ID = ATMF VOD
Generic identifier transport		H.310 correlation ID		H.310 Correlation ID DSM-CC Session/Resource ID

signaling exchanges need to support classical IP over ATM implementations and are specified in RFC 1577. The implementer's guide, RFC 1755, provides interoperability among implementations of RFC 1577, RFC 1483, and UNI ATM signaling. ATM signaling provides indication and negotiation between ATM endpoints for selection of end-to-end protocols and their parameters.

The values of various information elements used to characterize the IP over ATM (i.e., classical IP) service are defined in Table 2.18.

Table 2.18
Q.2931 Signaling IEs Used in IP Over ATM Framework

Information Element	Service type = UBR	Service type = ABR
AAL parameters	AAL Type = AAL-5 Forward CPCS SDU size—desired IP MTU Backward CPCS SDU size—desired IP MTU SSCS type = Null	
ATM traffic descriptor	PCR line rate Best-effort indicator Frame discard indicator	PCR desired value MCR desired value Frame discard indicator
Minimum acceptable ATM traffic descriptor	PCR—desired value	
ABR setup parameters		ABR setup parameter values
Broadband bearer capability	Bearer class = BCOB-X Susceptibility to clipping = 0 User plane connection = PtP	Bearer class = BCOB-X Broadband transfer capability = 0 × 0C Susceptibility to clipping = 0 User plane connection = PtP
Broadband low-layer information	For indicating LLC/SNAP encapsulation User information layer 2 = LAN LLC (ISO 8802/2) For indicating VC use by IP only User information layer 2 = LAN LLC (ISO 8802/2) User information layer 3 = ISO TR 9577 with IPI = IP	
QoS parameter	QoS class forward : 0 QoS class backward: 0	
Calling and called party number	ATM endsystem address Type of number = unknown Addressing/numbering plan = ATM endsystem address	

3

UNI Point-to-Multipoint Signaling

A point-to-multipoint connection allows one end station to send its traffic to two or more end stations (see Figure 3.1). The station that generates the traffic is referred to as the root of the connection, whereas an end station that receives this traffic is referred to as a leaf. As the traffic can only flow from the root to the leaves, a point-to-multipoint connection is always unidirectional. In particular, a leaf can not use a point-to-multipoint connection to send traffic to the root or to any other leaf. This chapter is based on ITU-T Recommendation Q.2971, which specifies the signaling protocol to establish, maintain, and clear point-to-multipoint virtual channel calls/connections at the user-network interface.

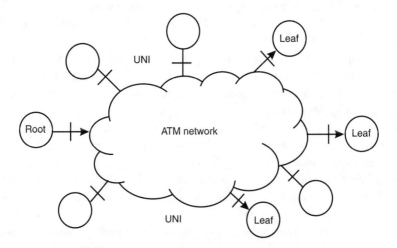

Figure 3.1 A point-to-multipoint connection at the UNI.

117

Let us consider the broadcast/unknown server (BUS) used in the ATM Forum LAN Emulation service. This service is defined to emulate the broadcast feature of shared media-based LANs such as Ethernet and Token Ring. In these LANs, once a packet is placed on the medium shared among multiple end stations, the packet is received by every station on that LAN. Each station filters the destination address placed in the packet and may copy the packet if the address matches one of the addresses configured in that station. One special such address is the broadcast address, which causes every station in the LAN to copy a packet with the broadcast address. Protocols designed and developed to run over these LANs take advantage of this broadcast capability and use it, for example, to resolve higher layer protocol addresses to medium access control (MAC)-level addresses. The LAN Emulation service in ATM is designed to run these applications over switched ATM networks without any changes, with BUS providing this broadcast feature. Every ATM end station that participates in the LAN Emulation service establishes a point-to-point connection to the BUS. If a packet is to be broadcast, the end station sends the packet (in cells) using its connection to the BUS, and the BUS forwards the cells to all end stations that participate in a particular emulated LAN. If there are N stations participating in the emulated LAN, one way to broadcast received cells to all end stations is to send it to each end station using point-to-point connections. In this case, a received cell needs to be copied N times (or $N-1$ times, excluding the end station that originated the cell). Alternatively, if a point-to-multipoint connection is established to all N end stations, there is no need to copy the received cell. The BUS transmits a single copy of the cell and the copying function is performed in the network at different switches and delivered to all N stations.

Point-to-multipoint connections provide the basic framework for utilizing network resources more efficiently. In particular, N cells are carried in the network if the end stations are reached via point-to-point connections. Using point-to-multipoint connections, the cell copy function may take place much closer to the end stations, thereby using fewer network resources. The effective utilization of network resources may be more important with high-bandwidth multimedia applications. For example, when a movie is distributed from a video server to N subscribers, the total bandwidth used in the network to reach N subscribers may be significantly more with point-to-point connections than a point-to-multipoint connection.

Using point-to-multipoint connections to reach multiple end stations also reduces the number of VPI/VCIs used both in the end station and in switches. In particular, each point-to-point connection needs to be uniquely identified at each interface the connection passes through between the source and the destination end station whereas a point-to-multipoint connection requires a single connection identifier at the source UNI and at each NNI it passes through in the network.

A point-to-multipoint connection is set up by first establishing a point-to-point unidirectional connection from the root (the calling party) to a leaf (a called party) using point-to-point signaling procedures. It is necessary to know that a connection is a point-to-multipoint connection before it is signaled to the network. In particular, it is not possible to add a third party to a connection after it is set up as a bidirectional point-to-point connection.

After the first setup to a leaf has been completed or the leaf is alerted and the root receives the corresponding indication, the root can request that new leaves be added to the connection, either one at a time or there may be more than one request pending simultaneously. That is, a root doesn't need to wait for a response to an add party request before issuing the next one. However, each request is exclusively related to one called party. Leaves can be added or dropped from the call at any time while the call is in the *active* state. To be a leaf of a point-to-multipoint connection, it is not necessary for a leaf to support point-to-multipoint signaling procedures. The procedures are defined so that if an end station supports Q.2931 (i.e., point-to-point) signaling messages, it can become a leaf of a point-to-multipoint call.

3.1 An Overview of Point-to-Multipoint Signaling Framework

The calling party is required to know ahead of time that its connection request is point-to-multipoint and the connection setup to the first party is required to have an indicator characterizing the connection as point-to-multipoint. The point-to-multipoint signaling procedures are used at a UNI after the first party is set up across that interface.

The point-to-multipoint signaling framework at a UNI includes the specifications of signaling messages, information elements, and procedures used to manage point-to-multipoint connections across a UNI.

3.1.1 An Overview of Point-to-Multipoint Signaling Messages

The formats of messages and the information elements used in point-to-multipoint signaling are the same as they are in point-to-point signaling. In order to communicate and manage signaling relationships with multiple UNIs, two new information elements are defined: *endpoint reference* IEs and *endpoint state* IEs, as illustrated in Figures 3.2 and 3.3. An endpoint reference IE is added to ALERTING, CALL PROCEEDING, CONNECT, SETUP, STATUS, STATUS ENQUIRY, and NOTIFY point-to-point signaling messages to

Information Element	Direction	Type	Length
Endpoint reference	both	O	4–7

Figure 3.2 Endpoint reference IE used in Q.2931 signaling messages.

Information Element	Direction	Type	Length
Endpoint reference	both	O	4–7
Endpoint state	both	O	4–5

Figure 3.3 Information elements used in the STATUS message for point-to-multipoint connections.

uniquely identify at the local UNI the leaf to which the signaling message corresponds. The endpoint state IE is included in the STATUS message when the endpoint reference IE is included, and it is used to indicate the call state (i.e., party state) of the leaf.

These additional IEs are mandatory in the above list of signaling messages if they were included in the SETUP message, and they are mandatory IEs in the SETUP message if the connection is point-to-multipoint.

After the point-to-point connection of a point-to-multipoint connection enters the active state (i.e., it is established) or the *alerting* state (i.e., the called party has alerted) at a UNI, point-to-multipoint signaling messages are used to add new leaves and to clear the connection to the existing leaves. The signaling messages used to manage point-to-multipoint calls are listed in Table 3.1.

Table 3.1
Point-to-Multipoint Signaling Messages

Message	Description
ADD PARTY	Used to add a new leaf to an already established point-to-multipoint connection
ADD PARTY ACKNOWLEDGE	Sent from the network to the calling party (root) or from the called party (leaf) to the network to acknowledge that the ADD PARTY message for a particular leaf was successful
PARTY ALERTING	Sent from the network to the calling party notifying that party alerting for a particular leaf has been initiated

Message	Description
ADD PARTY REJECT	Sent from the network to the root or from a leaf to the network that connection can not be extended to the leaf
DROP PARTY	Sent either by the user or by the network to request the connection to a particular leaf be cleared
DROP PARTY ACKNOWLEDGE	Sent either by the user or by the network to acknowledge that the connection to a particular leaf is successfully cleared

3.1.2 Point-to-Multipoint Party States

The call states of point-to-point signaling procedures are discussed in Chapter 2. Once the call is in an alerting state or in an active state, the add party or drop party procedures can be initiated. The states that a call goes through when a point-to-point connection to a leaf is set up are referred to as the link states (as opposed to call states used in regular point-to-point signaling procedures) to differentiate them from their counterparts used in point-to-point signaling.

The states a call goes through during point-to-multipoint signaling procedures are referred to as party states. These states may exist both at the user side and the network side of the UNI. The party states defined in Q.2971 are listed in Table 3.2.

Table 3.2
Party States

Party State	Description
Null (P0)	The party does not exist
Add Party Initiated (P1)	A SETUP or an ADD PARTY message has been sent to the other side of the interface for this party for the call
Add Party Received (P2)	A SETUP or an ADD PARTY message has been received from the other side of the interface for this party for the call
Party Alerting Delivered (P3)	An ALERTING or a PARTY ALERTING message has been sent to the other side of the interface for this party for the call
Party Alerting Received (P4)	An ALERTING or a PARTY ALERTING message has been received from the other side of the interface for this party for the call

Table 3.2 (Continued)

Party State	Description
Drop Party Initiated (P5)	A DROP PARTY message has been sent for this party of the call
Drop Party Received (P6)	A DROP PARTY message has been received for this party of the call
Active (P7)	On the user side of the UNI, when the user has received a CONNECT, CONNECT ACKNOWLEDGE, or ADD PARTY ACKNOWLEDGE message identifying this party, or sent an ADD PARTY ACKNOWLEDGE On the network side of the UNI, when the network has sent a CONNECT, CONNECT ACKNOWLEDGE, or ADD PARTY ACKNOWLEDGE message identifying this party, or when the network has received an ADD PARTY ACKNOWLEDGE message identifying this party

Each leaf is uniquely identified across an interface by assigning a unique endpoint reference value to each reachable party. Both the link state and the party states of each reachable party are maintained at each side of the UNI the root communicates with the network. If only one end station (leaf) is attached to the interface, only the link state needs to be maintained at the UNI. Other than the UNI the root is attached to at the network, the only other UNI at which both the link state and party states are maintained at each side of the UNI may be the interface between a public and private network. This is the case if the root is attached to a public (private) network and there are one or more leaves attached to a private (public) network.

3.2 Point-to-Multipoint Control Messages

This section presents the details of point-to-multipoint signaling messages.

3.2.1 ADD PARTY

The ADD PARTY message is sent from the calling party (root) to the network to request the addition of a party to an already established point-to-multipoint connection (see Figure 3.4). It is also used in the network in the called user direction to extend the existing connection to a new party. That is, this message

Information Element	Direction	Type	Length
AAL parameters	both	O	4–21
Broadband high-layer information	both	O	4–3
Broadband low-layer information	both	O	4–17
Called party number	both	M	≤ 25
Called party subaddress	both	O	4–25
Calling party number	both	O	≤ 25
Calling party subaddress	both	O	4–25
Transit network selection	U->N	O	4–8
Endpoint reference	both	M	7
End-to-end transit delay	both	O	4–10

Figure 3.4 Information elements used in an ADD PARTY message.

is used at a UNI only if a setup to at least one other party has already progressed at the interface.

The following IEs are included in the user-to-network direction when the calling user wants to pass the corresponding information to the called user.

- Calling party subaddress;
- AAL parameters;
- Broadband high-layer information;
- Broadband low-layer information;
- Called party subaddress.

One or more of these IEs are included in the network-to-user direction if the calling user placed the IE in the ADD PARTY message. The contents of these IEs are required to have the same values as in the initial SETUP message of the call. However, their contents are not checked by the network.

In addition to these, the ADD PARTY message may include the following IEs. The *called party* IE is included by the calling user to identify the destination end station. The *calling party number* IE may be included by the calling user or by the network to identify the calling user. The *transit network selection* IE is included by the calling user to select a particular transit network. The *end-to-end transit delay* IE is included in the user-to-network direction when the calling user wants to specify the end-to-end transit delay requirement to the called party. It is included in the network-to-user direction if the end-to-end transit delay information is to be delivered to the called user. The *endpoint reference* IE is used to uniquely identify the leaf to which the signaling message corresponds.

3.2.2 ADD PARTY ACKNOWLEDGE

The ADD PARTY ACKNOWLEDGE message is sent by the network to the user to acknowledge that the corresponding ADD PARTY request uniquely identified by the endpoint reference value was successful (see Figure 3.5). Similarly, it is sent in the user-to-network direction to acknowledge that the ADD PARTY request was successful.

The *AAL parameters* and the *broadband low-layer information* IEs may be included when a point-to-multipoint connection is established with an end station that does not support the point-to-multipoint signaling procedures (i.e., across a public UNI). The end-to-end transit delay IE is included in the user-to-network direction when the responding user received this IE in the SETUP or ADD PARTY message. It is included in the network-to-user direction if the responding user included it in the CONNECT or ADD PARTY ACKNOWLEDGE message. The endpoint reference IE is used to uniquely identify the leaf to which the signaling message corresponds.

3.2.3 PARTY ALERTING

The PARTY ALERTING message is sent by the network to the user to indicate that the called party alerting has been initiated (see Figure 3.6). It is also used in the user-to-network direction to indicate that the called party alerting has been initiated.

The endpoint reference IE is used to uniquely identify the leaf to which the signaling message corresponds.

Information Element	Direction	Type	Length
Endpoint reference	both	M	7
AAL parameters	both	O	4–21
Broadband low-layer information	both	O	4–17
End-to-end transit delay	both	O	4–10

Figure 3.5 Information elements used in an ADD PARTY ACKNOWLEDGE message.

Information Element	Direction	Type	Length
Endpoint reference	both	M	7

Figure 3.6 Information element used in a PARTY ALERTING message.

3.2.4 ADD PARTY REJECT

The ADD PARTY REJECT message is sent by the network to the user to indicate that the ADD PARTY request was not successful (see Figure 3.7). It is also used in the user-to-network direction when the ADD PARTY request is not successful.

The endpoint reference IE is used to uniquely identify the leaf to which the signaling message corresponds. The *cause* IE is mandatory in the ADD PARTY REJECT message to indicate the reason for rejecting the request.

3.2.5 DROP PARTY

The DROP PARTY message is sent by the root to the network or by the network to the root to drop a party from an existing point-to-multipoint connection (see Figure 3.8). It is also used in both directions at an interface (such as a UNI between a private ATM network and a public ATM network) from which one or more leaves are reached to drop a party from an existing point-to-multipoint connection.

The endpoint reference IE is used to uniquely identify the leaf to which the signaling message corresponds. The cause IE is mandatory in the DROP PARTY message to indicate the reason for dropping a party.

3.2.6 DROP PARTY ACKNOWLEDGE

The DROP PARTY ACKNOWLEDGE message is sent by the root to the network or by the network to the root in response to a DROP PARTY message to indicate that the party was dropped from the connection (see Figure 3.9). It is

Information Element	Direction	Type	Length
Cause	both	M	6–34
Endpoint reference	both	M	7

Figure 3.7 Information elements used in an ADD PARTY REJECT message.

Information Element	Direction	Type	Length
Cause	both	M	6–34
Endpoint reference	both	M	7

Figure 3.8 Information elements used in the DROP PARTY message.

Information Element	Direction	Type	Length
Cause	both	O	4–34
Endpoint reference	both	M	7

Figure 3.9 Information elements used in the DROP PARTY ACKNOWLEDGE message.

also used in both directions at an interface (such as a UNI between a private ATM network and a public ATM network) from which one or more leaves are reached to drop a party from an existing point-to-multipoint connection.

The cause IE is mandatory in the first party dropping message when the DROP PARTY ACKNOWLEDGE is sent as a result of an error-handling condition to specify the nature of the error. This IE may appear twice in this message. This happens when the DROP PARTY ACKNOWLEDGE message is sent by the party who originally sent the DROP PARTY message but did not receive a DROP PARTY ACKNOWLEDGE message from the other party before its timer expired. In this case, the party DROP PARTY ACKNOWLEDGE message contains the original cause sent in the DROP PARTY message and a second cause that resulted in sending the DROP PARTY ACKNOWLEDGE message (i.e., timer expired).

3.3   Information Elements Used in Point-to-Multipoint Signaling Messages

The IEs and coding rules defined in Q.2931 for point-to-point calls/connections also apply to the IEs used in point-to-multipoint signaling procedures. Two additional IEs are defined for the point-to-multipoint signaling procedures to work properly.

3.3.1   Endpoint Reference Information Element

The purpose of the endpoint reference IE is to identify the individual end points of a point-to-multipoint call to which the particular message applies (see Figure 3.10).

The only endpoint reference type currently defined is the *locally defined integer*. All other values are reserved. The endpoint reference flag is a 1-bit field. If it is set to 0, the message is sent from the side that originated the endpoint reference, whereas the message is sent to the side that originated the endpoint reference if it is set to 1. The endpoint reference identifier is a 15-bit integer (coded in binary) to uniquely identify an end point at a UNI. The value of 0 is reserved to identify the first party of a point-to-multipoint call. This is because

Endpoint Reference Type		5
0/1 Endpoint reference flag	Endpoint reference identifier value	6
	Endpoint reference identifier value (continued)	6.1

Figure 3.10 Endpoint reference IE.

the first party can negotiate the connection parameters using the point-to-point signaling negotiation procedures. A nonzero endpoint reference value is always used to identify the subsequent parties of a point-to-multipoint call. Subsequently called parties cannot negotiate the connection characteristics and they are required to either accept the parameters agreed upon during the connection establishment to the first party or reject the add party request.

3.3.2 Endpoint State Information Element

The purpose of the endpoint state IE is to indicate the state of an end point of a point-to-multipoint connection (see Figure 3.11).

The currently defined endpoint reference party states in Q.2971 are as follows:

- Null;
- Add Party Initiated;
- Party Alerting Delivered;
- Add Party Received;
- Party Alerting Received;
- Drop Party Initiated;
- Drop Party Received;
- Active.

3.4 Point-to-Multipoint Signaling Procedures

This section describes the point-to-multipoint call management procedures at the UNI. The corresponding signaling messages are exchanged using the signal-

0 0 Spare	Endpoint reference party state	5

Figure 3.11 Endpoint state IE.

ing virtual channel VPCI = 0, VCI = 5—the same one used for point-to-point calls. Point-to-multipoint signaling messages use the Q.2931 protocol discriminator value, included in every message. These procedures also include the capability for a leaf to drop itself from a point-to-multipoint call. These signaling procedures apply after the initial point-to-point procedures are completed.

3.4.1 Connection Setup to the First Party

The first connection to a leaf is set up using the Q.2931 procedures as described in Chapter 2. The SETUP message sent by the root is required to contain both the endpoint reference IE and the *broadband bearer capability* IE with the indication of point-to-multipoint in the user plane connection configuration field. Since a point-to-multipoint connection is always unidirectional, the backward cell rate in the ATM traffic descriptor IE is set to 0. If a nonzero value is specified, the network rejects the set up request with cause "unsupported combination of traffic parameters." OAM flows are not supported for point-to-multipoint connections. Accordingly, the OAM traffic descriptor IE is not included in the SETUP message.

This SETUP message contains an endpoint reference IE to uniquely identify the leaf to which the message applies. If this IE is missing, the network rejects the setup request with cause "mandatory information element is missing" and includes the endpoint reference IE identifier in the diagnostic field.

Point-to-point signaling allows the negotiation of both traffic and QoS parameters of a call during the call setup time. In the case of point-to-multipoint signaling, this capability is allowed only during the call setup to the first leaf. If the negotiation is initiated, the root is required to set the endpoint reference value to 0 in the SETUP message and not send any ADD PARTY messages until it finds out that the called party is alerted or has accepted the connection. This is required to guarantee the completion of the negotiation process. Once the corresponding call parameters are agreed upon among the root, the first leaf, and the network, the call setup requests to all additional leaves use the same negotiated parameters.

Upon sending a SETUP message to the network, the root enters the *add party initiated* party state. For the connection setup to the first party, the party state timer T399 is not started. The network may respond to this setup request by sending a CALL PROCEEDING message (sending this message does not cause any party state changes). When the network receives a SETUP message, it enters the *add party received* party state. If the network sends an ALERTING message, it enters the *party alerting delivered* party state. Upon receiving an ALERTING message, the calling party enters the *party alerting received* state and starts timer T397 (if implemented). In this case, timer T301 is not started.

If the call is accepted by the network and the called party, the network sends a CONNECT message to the root. Upon receiving this message, the calling party stops timer T310 or T397 and enters the active party state. As a response to the CONNECT message, the recipient sends a CONNECT ACKNOWLEDGE message. Sending this message does not cause the party state to change. Similarly, sending a CALL PROCEEDING message or receiving a CONNECT ACKNOWLEDGE message by the network does not result in any changes in the party state.

If an ALERTING, CALL PROCEEDING, or CONNECT message is received by the user in response to a SETUP message with a missing endpoint reference IE, or if this IE has a content error, the Q.2931 mandatory information element error procedures apply.

3.4.2 Adding a Party at the Originating Interface

A new leaf is added to a point-to-multipoint connection by sending an ADD PARTY message. An ADD PARTY message can be sent only if the link is in the active or call delivered link state. Each ADD PARTY message has the same call reference value as specified in the initial setup of the point-to-multipoint call. Upon sending an ADD PARTY message, the root starts timer T399 and enters the add party initiated party state.

Upon receiving the ADD PARTY message, the network enters the add party received party state. If the network decides to accept the request, it progresses the call to the destination UNI.

After receiving an indication that called party alerting has been initiated, the network sends a PARTY ALERTING message to the root and enters the party alerting delivered party state for that party. Upon receiving the PARTY ALERTING message, the user stops timer T399, starts timer T397 (if implemented), and enters the party alerting received party state. Figure 3.12 illustrates the process of adding a subsequent party at the originating interface.

The state transition diagram and the sequence of events in sending and receiving an ADD PARTY message are illustrated in Figures 3.13 to 3.15.

If timer T397 (if implemented) or T399 expires, the user initiates party dropping procedures with cause "recovery on timer expiry" according to the root-initiated party dropping procedures.

After the add party request is accepted, the network sends an ADD PARTY ACKNOWLEDGE or CONNECT message to the root and enters the active party state for that party. Either message indicates to the root that an additional party has been added to the original connection. The CONNECT message is used if the call is in the call delivered link state, while the ADD PARTY ACKNOWLEDGE message is used if the call is in the active link state.

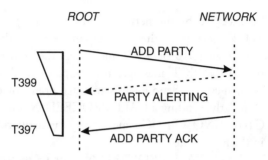

Figure 3.12 Adding a subsequent party at the originating interface.

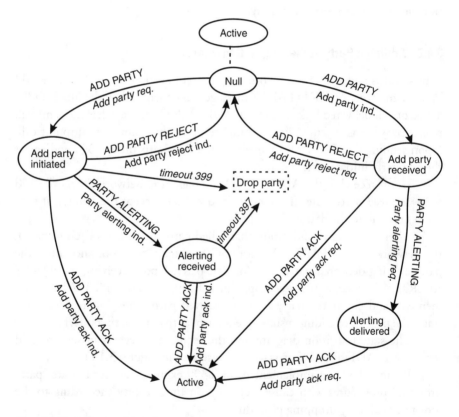

Figure 3.13 State transition diagram.

Upon receiving an ADD PARTY ACKNOWLEDGE or CONNECT message, the root enters the active party state and stops timer T399 (if running) or timer T397 (if implemented and running), as shown in Figure 3.16.

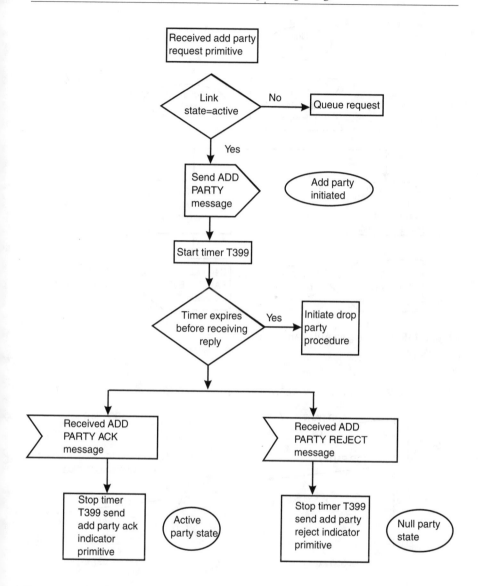

Figure 3.14 Sending an ADD PARTY message.

3.4.3 Add Party Rejection

If the network or the called user is unable to accept the call, the network rejects the party request with the cause provided by the network or the called user by sending an ADD PARTY REJECT message with one of the following causes: "user cell rate not available," "resource unavailable, unspecified," "quality of

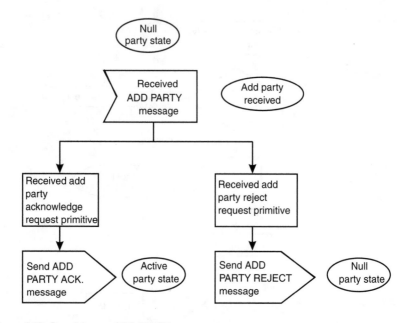

Figure 3.15 Receiving an ADD PARTY message.

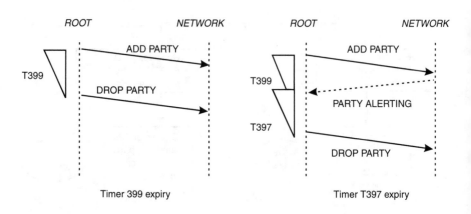

Figure 3.16 Timer expiry during add party procedures.

service unavailable," or "bearer capability not presently available." Similarly, if the call information received from the calling user is not valid, the network sends an ADD PARTY REJECT message with one of the following cause values: "unassigned (unallocated) number," "no route to destination," "number changed," or "invalid number format (address incomplete)."

3.4.4   Add Party Establishment at the Destination Interface

The addition of a party at the destination UNI is initiated by the network using the Q.2931 procedures with the following modifications. The network indicates the arrival of an add party request at the UNI by transferring a SETUP message across the interface. This message contains the endpoint reference and broadband bearer capability IEs indicating point-to-multipoint in the user plane connection configuration field. The instruction indicator in the endpoint reference IE is coded to "discard information element and proceed" for end stations that do not support the point-to-multipoint signaling procedures. The endpoint reference is set to 0 if it was 0 at the originating interface, indicating that this is the first party of the call, and the root is allowed to negotiate the call traffic and QoS parameters. In any other case, the endpoint reference value is nonzero.

For each party in a call, there is a unique endpoint reference value at each UNI. Messages that are exchanged at a UNI for a party contain the same call reference and the endpoint reference values specified in the SETUP message. If, however, a CALL PROCEEDING, ALERTING, or CONNECT message is the first message the network receives in response to its SETUP message and the message does not contain an endpoint reference IE, it is not treated as an error. This may happen when an end station does not support point-to-multipoint signaling procedures to be a leaf. If any of these messages is received with an endpoint reference IE with content error (e.g., incorrect endpoint reference value or flag), it is treated as a mandatory IE content error. Similarly, if the first message in response to the SETUP message is received with a correct endpoint reference and a subsequent ALERTING or CONNECT message is received with either no endpoint reference or with a content error, it is treated as a mandatory IE error.

After transmitting a SETUP message, the network enters the add party initiated party state while the called party enters the add party received party state upon receiving a SETUP message from the network with a point-to-multipoint indicator. The called party may respond with a CALL PROCEEDING message. Sending this message does not cause the party state at the called party to change. Similarly, receiving a CALL PROCEEDING message does not cause the network to change its party state. The called party may respond by transmitting an ALERTING message, which causes it to enter the party alerting delivered party state. Upon receiving an ALERTING message, the network enters the party alerting received party state.

After accepting the connection, the called party sends a CONNECT message. This does not cause the user to change its party state. If the leaf provides a broadband low-layer information IE or an AAL parameters IE in the CON-

NECT message, the network is required to deliver this IE to the root in a CON-
NECT or ADD PARTY ACKNOWLEDGE message, as appropriate, provided
that the network supports the delivery of the broadband low-layer information
IE.

When the network receives a CONNECT message, it enters the active
party state and sends back a CONNECT ACKNOWLEDGE message. This
message causes the user to enter the active party state. Upon sending a CON-
NECT ACKNOWLEDGE message, the network does not change the party
state.

At the destination interface, the ADD PARTY, ADD PARTY AC-
KNOWLEDGE, ADD PARTY REJECT, PARTY ALERTING, DROP
PARTY, and DROP PARTY messages are not used. If any one of these messages is received at the destination interface, it is treated as an unrecognized or
unexpected message. Figure 3.17 illustrates the process of adding a new leaf.

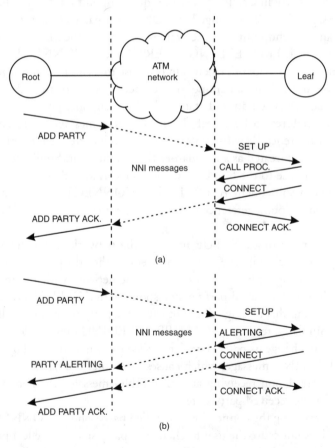

Figure 3.17 Adding a new leaf (a) without and (b) with ALERTING message.

The network may detect that the called party does not support point-to-multipoint signaling procedures if it receives a CALL PROCEEDING, ALERTING, or CONNECT message as the first response to a SETUP message at the UNI to which the called party is attached without an endpoint reference. Hence, when the response to the SETUP message does not include the endpoint reference IE, the network does not treat that as an error. Once it detects that the called party does not support the point-to-multipoint signaling procedures, the network may not include the endpoint reference or endpoint state IE in its subsequent messages. Alternatively, it may include this IE after coding the IE instruction field to "discard information element and proceed."

After the network detects that the called party does not support point-to-multipoint signaling procedures in the first response to its SETUP message and receives a subsequent signaling message with an endpoint reference IE, it does not treat that as an error. This is because there may be more than one leaf reached from this UNI (i.e., across a public UNI to which a private network is attached), and while some end stations attached to the public network may support point-to-multipoint signaling procedures, others may not. Finally, although a leaf may not include the endpoint reference IE at the destination interface, the inclusion of this IE is mandatory at the originating interface for the root to associate the message with the corresponding leaf.

In cases in which the user fully supports point-to-multipoint (i.e., a private network node), the network may add additional parties by sending ADD PARTY messages (Section 3.1).

3.4.5 Party Dropping

Under normal conditions, dropping a party is initiated when the user or the network sends either a DROP PARTY or RELEASE message. One exception to this rule occurs when the user or network responds to a SETUP message with a RELEASE COMPLETE message to reject the call (e.g., due to unavailability of a virtual channel, user cell rate is not available, etc.). In this case, Q.2931 call-clearing procedures are applied. Another exception is when the network sends an ADD PARTY REJECT message as the first response to an ADD PARTY message. In this case, the network enters the null party state. In addition, if there is no other party active or being established at the interface, then the network sends a RELEASE message with cause "normal unspecified."

When a leaf wants to leave a point-to-multipoint connection, it sends a RELEASE message and enters the null party state.

A root may initiate party dropping by sending a DROP PARTY or RELEASE message. The DROP PARTY message is used when the party being dropped and at least one other party are in the active, party alerting received or add party initiated party state. The RELEASE message is used, instead, if this is the last party of the call.

Throughout the rest of this chapter, the call control procedures are classified into two categories based on the link and party states of a call at a UNI. In the first category, in addition to the leaf for which the signaling message is sent, there is at least one other party at the interface in the active, party alerting delivered, or add party received party state. For the simplicity of the presentation, this state will be referred to as the *busy state*. Similarly, if all the other parties associated with the call are in the null, drop party initiated, or drop party received party state, the state will be referred to as the *idle state*.

After sending a DROP PARTY message, the root starts timer T398 and enters the drop party initiated party state.

When the network receives this message, it enters the drop party received party state. If the call is in a busy state, the network initiates procedures for dropping the party along the path to the remote user, sends a DROP PARTY ACKNOWLEDGE message to the root, and enters the null party state. If the call is in an idle state, the network enters the null party state, initiate procedures for dropping the party along the path to the remote user, and sends a RELEASE message to the root with cause "normal, unspecified." The DROP PARTY ACKNOWLEDGE message has a local significance only and it does not acknowledge that the leaf node is dropped from the call.

Upon receiving a DROP PARTY ACKNOWLEDGE message, the root stops timer T398 and returns to the null party state. If the call is in an idle state, the user releases the call by sending a RELEASE message with cause "normal unspecified."

If timer T398 expires and if the call is in a busy state, the user sends a DROP PARTY ACKNOWLEDGE message to the network with the cause number originally contained in the DROP PARTY message and enters the null party state. In addition, the user may indicate a second cause IE with cause "recovery on timer expiry." If the call is in an idle state when the timer T398 expires, the user sends a RELEASE message to the network with the cause number originally contained in the DROP PARTY message. In addition, the user may indicate a second cause IE with cause "recovery on timer expiry." Figure 3.18 illustrates the process of dropping a party.

When the network receives a RELEASE message, any party in the drop party initiated or drop party received party state enters the null party state and all other parties are dropped towards the leaves, with the cause contained in the RELEASE message or cause "normal unspecified" if no clause was included in the RELEASE message. At the destination UNI, the network initiates party dropping with a RELEASE message.

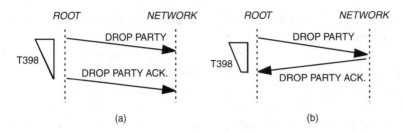

Figure 3.18 Dropping a party: (a) timer T398 expiry and (b) root-initiated drop party.

When party dropping is initiated by the network, a DROP PARTY or RELEASE message is sent at the root interface. A DROP PARTY message is used to initiate party dropping when the party is in the active or party alerting delivered party states and the call is in a busy state. Otherwise, the RELEASE message is sent. The party dropping procedures by the network are the same as

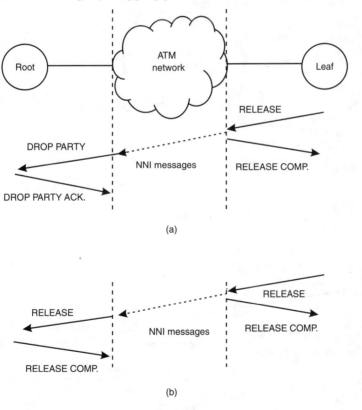

Figure 3.19 Leaf-initiated party dropping: (a) not the last party of the call/connection and (b) the last party of the call/connection.

the root-initiated dropping procedures. Figure 3.19 illustrates the process of leaf-initiated party dropping.

If a DROP PARTY or ADD PARTY REJECT message is received while the party is in the drop party initiated party state and the call is in a busy state, the recipient stops timer T398 and enters the null party state. Similarly, if a DROP PARTY or ADD PARTY REJECT message is received when the party is in the drop party initiated party state and there is no other party active or in party alerted state, the recipient stops timer T398, disconnects the bearer virtual channel, and sends a RELEASE message.

It is also possible for a DROP PARTY ACKNOWLEDGE message to cross with a RELEASE message at the UNI. The network ignores the DROP PARTY ACKNOWLEDGE message if it is in the release indication link state. Similarly, the user ignores DROP PARTY ACKNOWLEDGE message while it

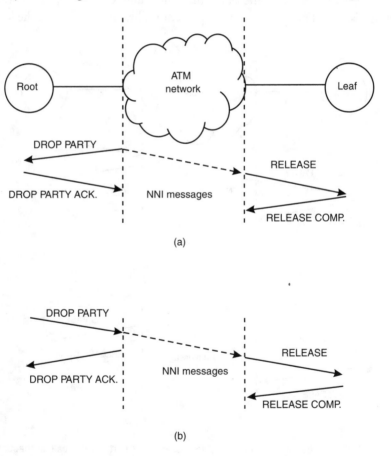

Figure 3.20 (a) Network- and (b) root-initiated party dropping (not the last party of the call/connection.

is in the release request link state. Figure 3.20 illustrates the network- and root-initiated party dropping process.

The above procedures apply when a single party is dropped. If the root wants to drop all parties of a call, it sends a RELEASE message to the network. In this case, the network first sends an ADD PARTY REJECT message for each party in the add party received party state and then sends a RELEASE message.

The state transition diagram for dropping a party is illustrated in Figure 3.21.

3.5 Restart Procedure

When a virtual channel is restarted, the user and the network are required to drop all parties associated with the virtual channel. The network initiates normal drop party procedures toward the leaves for all parties associated with the call, and related Q.2931 procedures apply.

3.6 Handling of Error Conditions

The error-handling procedures in point-to-multipoint signaling for the protocol discrimination, message too short, message length, and general information element errors are the same as point-to-point error-handling procedures discussed in Chapter 2. In this section, we discuss error-handling procedures that apply specifically to adding or dropping parties in a point-to-multipoint call.

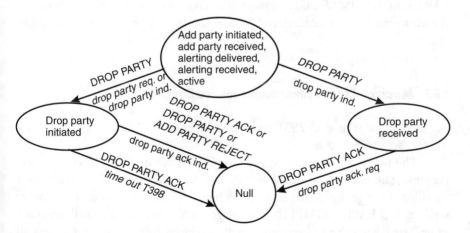

Figure 3.21 Party dropping state transitions.

3.6.1 Endpoint Reference Information Element Related Errors

Point-to-multipoint signaling procedures use the endpoint reference IE to identify the leaf to which the message corresponds. Two error conditions are identified with regards to this IE: (1) missing endpoint reference IE and (2) invalid endpoint reference format. The error-handling procedures defined to handle these errors specify what happens when the receiving entity cannot determine the leaf to which the received message corresponds.

If an ADD PARTY REJECT, DROP PARTY, or DROP PARTY ACKNOWLEDGE message is received without an endpoint reference IE, the receiving entity clears the call with cause "mandatory information element is missing." If the received message is ADD PARTY, PARTY ALERTING, or ADD PARTY ACKNOWLEDGE, the receiving entity checks the IE instruction field. If the IE instruction field indicates "follow explicit instructions and clear the call," the call is cleared with cause "invalid information element contents." If the IE instruction field indicates "follow explicit instructions and discard message and report status," a STATUS message is returned with cause "information element nonexistent or not implemented" or with cause "invalid information element contents." If the IE instruction field indicates "follow explicit instructions and discard message," the message is ignored. If none of these conditions apply, a STATUS message is returned with cause "invalid information element contents."

The received signaling message may include the endpoint reference IE, but an error might have occurred causing its contents to change (i.e., type is not equal to 0), or its length may be incorrect. When such an error is detected, the IE becomes invalid. When an ADD PARTY, PARTY ALERTING, or ADD PARTY ACKNOWLEDGE message is received with an invalid reference IE, the error-handling procedures are defined the same way as missing endpoint reference IE.

3.6.2 Message Type or Message Sequence Errors

In general, the related Q.2931 procedures apply to handle message type and message sequence errors.

The point-to-multipoint signaling messages are used at a UNI only after the link state is active or call delivered link state. If any point-to-multipoint signaling message is received at a time when the link state is null, the receiver sends a RELEASE COMPLETE message with cause "invalid call reference value," specifying the call reference in the received message and remaining in the null link state.

If the receiving entity is in the null party state, the only point-to-multipoint signaling message it expects to receive is the ADD PARTY message. If a DROP PARTY ACKNOWLEDGE or ADD PARTY REJECT message is received for a party in the null party state, no action is taken. If a CALL PROCEEDING, ALERTING, or CONNECT message is received, a RELEASE message is sent with cause "invalid endpoint reference value." If any other message is received for a party in the null party state, a DROP PARTY ACKNOWLEDGE message is sent with cause "invalid endpoint reference value" and the receiver remains in the null party state.

An ADD PARTY message is expected to arrive when the party is in the null party state. If this message is received and the party is in the add party received party state, the receiving entity ignores the message. If the party is in another state, a STATUS message is sent with the call state, the associated endpoint reference and endpoint state IEs with a cause "message not compatible with call state."

If a RELEASE COMPLETE message is received unexpectedly (i.e., the receiver did not originate the corresponding RELEASE message), it is treated as if a RELEASE message is received and the corresponding procedures for handling the party states are applied.

If the network receives a DROP PARTY ACKNOWLEDGE message unexpectedly and the call is in the active or call delivered link state, it initiates normal party dropping procedures toward all leaves with the cause indicated by the user, stops the timers set for each party, and enters the null party state. If the cause is not included, the cause "protocol error, unspecified" is used.

If a user receives an unexpected DROP PARTY ACKNOWLEDGE message from the network and the call is in the active or call delivered link state, it stops all party state timers and enters the null party state. If the call is not in a busy state at the time the DROP PARTY ACKNOWLEDGE message is received, the side that receives the DROP PARTY ACKNOWLEDGE disconnects the bearer virtual channel and sends a RELEASE message.

3.6.3 Mandatory Information Element Related Errors

The Q.2931 procedures apply, in general, if a mandatory IE is received and there is an error in the IE. If an ADD PARTY message is received with one or more mandatory IEs missing, the receiving entity sends an ADD PARTY REJECT message with cause "mandatory information element is missing." When a DROP PARTY ACKNOWLEDGE or ADD PARTY REJECT message is received without a cause IE, the receiver processes the message as if it was received with cause "normal, unspecified." If the cause IE is missing in a DROP PARTY message, the receivers treats this message as a DROP PARTY message with cause "normal, unspecified." When the DROP PARTY ACKNOWLEDGE or RELEASE message, as appropriate, is sent in response to this DROP PARTY message, it contains the cause "mandatory information element is missing."

If an ADD PARTY message is received with one or more mandatory IEs with invalid contents, an ADD PARTY REJECT or RELEASE message (as appropriate) with cause "invalid information element contents" is returned. If a DROP PARTY, DROP PARTY ACKNOWLEDGE, or an ADD PARTY REJECT message is received with invalid contents in the cause IE, the actions taken are the same as those taken when receiving the same message with the cause "normal, unspecified."

The Q.2931 procedures also apply, in general, if a nonmandatory IE is received and there is an error in the IE. If the received message is a DROP PARTY ACKNOWLEDGE or ADD PARTY REJECT message, no action is taken on the unrecognized information.

3.6.4 Status Enquiry Procedure

To verify a party state, the user or network sends a STATUS ENQUIRY message with the endpoint reference of the party for which it sends the request. Upon receiving a STATUS ENQUIRY message that includes an endpoint reference IE, the receiver responds with a STATUS message, reporting the current party state, the link state, and cause "response to status enquiry."

The corresponding Q.2931 procedures apply when a STATUS message that reports an incompatible link state is received. In this case, the receiving entity drops the party by sending the appropriate message with cause "message not compatible with call state."

If a STATUS message indicating any party state except the null party state is received in the null party state, the receiving entity sends a DROP PARTY ACKNOWLEDGE message with cause "message not compatible with call state," and remains in the null party state. If the received party state indicates null party, the receiver takes no actions and remains in the null party state.

If the party is in the drop party initiated party state and the STATUS message indicates any party state except the null party state, no action is taken.

If a STATUS message indicating that the party is in the null party state is received in any party state except the null party state, the receiver internally drops the party and enters the null party state. If no other party of the call is in the active, add party initiated, party alerting received, party alerting delivered, or add party received party states, then call clearing is initiated with a RELEASE message.

Rules for other types of exceptions to handle incompatible party states are left as implementation specific.

A STATUS message may be received indicating a compatible party state with one of the following causes: "mandatory information element is missing," "message type nonexistent or not implemented," "information element nonexistent or not implemented," or "invalid information element contents." In this case, the actions taken by the receiver are implementation specific. If no procedure is defined, the receiver drops the party using the cause specified in the received STATUS message.

Sending or receiving a STATUS message does not affect the party state of either the sender or the receiver. The receiver inspects the cause IE. If the cause is "response to STATUS ENQUIRY," timer T322 is stopped and the appropriate action is taken based on the information in the STATUS message and the current link state and party state of the receiver.

3.7 Procedures for Interworking With Private B-ISDNs

These procedures apply when the root of a point-to-multipoint connection is attached to a private UNI where one or more leaves of the call are attached to a public UNI and reached across a single public UNI, as illustrated in Figure 3.22.

The following procedures apply at the public UNI that connects the two networks to each other (hereafter referred to as public/private UNI for presentation purposes). In this context, the network refers to the public network and user refers to the private network. The point-to-multipoint signaling procedures used at the public/private UNI are slightly different from the procedures defined in Section 3.4. In order to reach to a party attached to a private network, the public network requests the establishment of a party across the public/

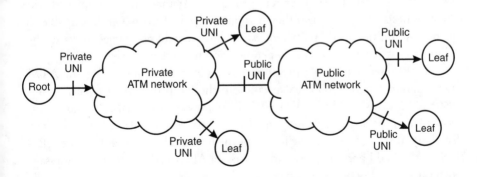

Figure 3.22 A point-to-multipoint call across a public UNI.

private UNI by transferring a SETUP or ADD PARTY message. To set up the initial party at the public/private interface, the point-to-point signaling procedures (with the indication in the SETUP message that this is the first party of a point-to-multipoint call/connection) apply with the following additions:

- Upon receiving an ALERTING message, the network starts timer T397 instead of timer T301.
- If a CALL PROCEEDING, ALERTING, or CONNECT message is the first message the network receives in response to the SETUP message and it does not contain an endpoint reference IE, it is not treated as an error.
- If timer T310 expires or timer T303 expires for the second time, the network initiates normal party drop procedures toward the root for all pending add party requests in the add party queue with cause "no user responding."

3.7.1 Incoming Add Party Request

Upon receiving and accepting an add party request, the network transfers an ADD PARTY message to the private network, starts timer T399, and enters the add party initiated party state. The ADD PARTY message is sent only if the link is in the active or call received link state. If the add party request is not accepted, it is dropped towards the root with cause "resource unavailable, unspecified." Upon receiving an ADD PARTY message, the user enters the add party received party state.

If there is one and only one party in the add party initiated party state and the call is not in the active or call received link state, additional add party requests are queued by the network until the link state becomes call received, active, null, release indication, or release request. At that time, the queued add party requests are treated as if they had just arrived. If the network is unable to queue any additional add party requests, it initiates add party reject towards the calling user with cause "too many pending add party requests."

If the user is not able to support the add party request with the ATM traffic parameters or QoS class specified in the setup request, it rejects the request and returns an ADD PARTY REJECT message with cause "resources unavailable, unspecified" or cause "quality of service unavailable," respectively. Similarly, if the user is not able to accept the indicated end-to-end transit delay, it rejects the add party request and returns an ADD PARTY REJECT message with cause "quality of service unavailable."

3.7.2 Response to an Add Party Request

Upon receiving an add party request, the user may respond with a PARTY ALERTING, ADD PARTY ACKNOWLEDGE, ADD PARTY REJECT, or CONNECT message, as appropriate (see Figure 3.23). If the add party request is accepted by the called party, the user sends an ADD PARTY ACKNOW-LEDGE or CONNECT message and enters the active party state. The CON-NECT message is used if the call is in the call received link state, whereas the ADD PARTY ACKNOWLEDGE message is used if it is in the active or connect request link states.

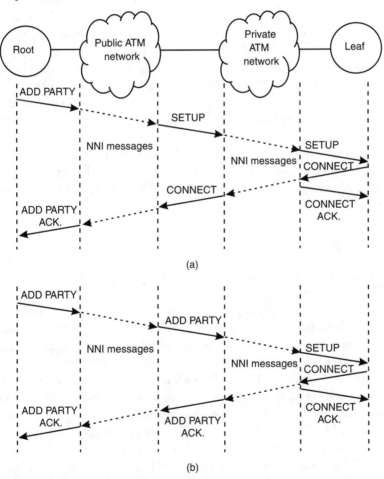

Figure 3.23 Adding a party: (a) first and (b) subsequent party at the public/private UNI.

When the network receives a PARTY ALERTING message from a private ATM switch, it stops timer T399, starts timer T397, enters the party alerting received party state, and sends a PARTY ALERTING message towards the root. If an add party request can be accepted and no user alerting is required, an ADD PARTY ACKNOWLEDGE or CONNECT message is sent.

If an ADD PARTY REJECT or DROP PARTY message is received when the party state is add party initiated or party alerting received, the network drops the party using party dropping features.

If the network does not receive any response to its ADD PARTY message until timer T399 expires, it initiates party dropping procedures towards the user with cause "no user responding." If this is the last party of the call, the network sends a RELEASE message to the called user with cause "normal unspecified."

After receiving an ALERTING or PARTY ALERTING message for the party, if the network does not receive a CONNECT, RELEASE, ADD PARTY ACKNOWLEDGE, or DROP PARTY message until timer T397 expires, it initiates party dropping towards the calling user with cause "no answer from user (user alerted)," and initiates party dropping at the destination interface with cause "recovery on timer expiry."

Upon receiving an ADD PARTY ACKNOWLEDGE or CONNECT message, the network stops timer T399 or T397, enters the active party state, and initiates procedures to send an ADD PARTY ACKNOWLEDGE message towards the root.

3.7.3 Party Dropping

Party dropping is initiated when the user or the network sends a DROP PARTY or RELEASE message.

A DROP PARTY message is used when the party is in the active, party alerting received, party alerting delivered, or add party initiated party state and there is at least one party of the call on this interface in the add party initiated, add party received, party alerting received, party alerting delivered, or active party state. After sending a DROP PARTY message, the user starts timer T398 and enters the drop party initiated party state. Upon receiving a DROP PARTY message, the network enters the drop party received party state.

The RELEASE message is used if all other parties belonging to the same call are in the null, drop party received, or drop party initiated party state. When a RELEASE message is used, the Q.2931 normal clearing procedures are used and all parties (for this call) are dropped. When the user receives a RELEASE message, any party in the drop party initiated, drop party received, add party received, party alerting delivered, party alerting received, or active party states party state enters the null party state.

3.8   Timers Used in Point-to-Multipoint Signaling

Various timers used in point-to-multipoint signaling procedures are summarized in Tables 3.3 and 3.4.

3.9   Handling the End-to-End Transit Delay Information Element

The use of the end-to-end transit delay IE for point-to-multipoint call/connection is mandatory for the network, while it is optional for the user. The purpose of this IE is to indicate the maximum end-to-end transit delay acceptable for an add party request and to indicate the expected cumulative transit delay to a leaf. The root may indicate a maximum end-to-end transit delay value to specify its delay requirements. Alternatively, it may indicate that any end-to-end transit delay is acceptable for the given party.

3.9.1   Originating UNI (SETUP or ADD PARTY)

The inclusion of the end-to-end transit delay IE in the SETUP or ADD PARTY message by the root is optional. If the root includes this IE in the SETUP or ADD PARTY message, both the cumulative transit delay and the maximum end-to-end transit delay subfields are present.

The user may set the maximum end-to-end transit delay subfield to "any end-to-end transit delay value acceptable, deliver cumulative end-to-end transit delay value" if any end-to-end transit delay is acceptable.

3.9.2   Destination UNI (SETUP or ADD PARTY)

The network includes an end-to-end transit delay IE if the root included it in the SETUP or ADD PARTY message for the party. Both the cumulative transit delay subfield and the maximum end-to-end transit delay subfield are present.

3.9.3   Called User

It is recommended that the called user update the cumulative transit delay value received from the network based on the status of its internal resources. If the cumulative transit delay value exceeds the maximum end-to-end transit delay value specified by the root, it is also recommended that the called user reject the add party request with cause "quality of service not available."

Table 3.3
Timers in the User Side

Timer	Default Time-out Value	Party State of Call	Cause for Start	Normal Stop	At the First Expiry	At the Second Expiry	Implementation
T397	Minimum of 3 min.	Party alerting received party state	ALERTING or ADD PARTY ALERTING received	ADD PARTY ACKNOWLEDGE received	Drop the party (DROP PARTY or RELEASE)	Timer is not restarted	Only mandatory if Annex A supported
T398	4 sec.	Drop party initiated party state	DROP PARTY sent	DROP PARTY ACKNOWLEDGE or RELEASE received	Send DROP PARTY ACKNOWLEDGE or RELEASE	Timer is not restarted	Mandatory
T399	34–124 sec.	Add party initiated party state	ADD PARTY sent	ADD PARTY ACKNOWLEDGE, PARTY ALERTING, or ADD PARTY REJECT received	Send DROP PARTY or RELEASE	Timer is not restarted	Mandatory

Table 3.4
Timers in the Network Side

Timer	Default Time-out Value	Party State of Call	Cause for Start	Normal Stop	At the First Expiry	At the Second Expiry	Implementation
T397	Minimum of 3 min.	Party alerting received party state	ALERTING or ADD PARTY ALERTING received	ADD PARTY ACKNOWLEDGE received	Drop the party (DROP PARTY or RELEASE)	Timer is not restarted	Mandatory
T398	4 sec.	Drop party initiated party state	DROP PARTY sent	DROP PARTY ACKNOWLEDGE or RELEASE received	Send DROP PARTY ACKNOWLEDGE or RELEASE	Timer is not restarted	Mandatory
T399	14 sec.	Add party initiated party state	ADD PARTY sent	ADD PARTY ACKNOWLEDGE, PARTY ALERTING, or ADD PARTY REJECT received	Send DROP PARTY or RELEASE	Timer is not restarted	Mandatory

3.9.4 Destination UNI (CONNECT or ADD PARTY ACKNOWLEDGE)

If the SETUP or ADD PARTY message sent to the called user included an end-to-end transit delay IE, the called user may include this IE in the CONNECT or ADD PARTY ACKNOWLEDGE message, specifying the final cumulative transit delay value for the add party request. The network does not check the correctness of the cumulative transit delay value provided by the user.

3.9.5 Originating UNI (CONNECT or ADD PARTY ACKNOWLEDGE)

The network includes an end-to-end transit delay IE in the CONNECT or ADD PARTY ACKNOWLEDGE message sent to the root if the called user included it in the CONNECT or ADD PARTY ACKNOWLEDGE message for the party.

3.10 Leaf-Initiated Join Capability

Point-to-multipoint signaling procedures discussed so far allow only one end station (the root) to add or delete leaves to a connection. The LIJ capability developed by ATM Forum allows users to independently join point-to-multipoint calls without intervention from the root of that connection. Moreover, leaves can request to join to a point-to-multipoint call independent of whether the call is active or inactive. ATM Forum UNI 4.0 specifies two types of LIJ calls: network leaf-initiated join and root leaf-initiated join.

In the *network LIJ*, the network is responsible for adding the leaves that request to join a call. The root is not notified when a leaf is added to or dropped from a connection (except when the root adds or drops the leaf itself, in which case the point-to-multipoint signaling procedures apply). It is possible that at the time the first leaf requested to join a call, the call may not be in an active state. In this case, the root is informed by the network to perform the initial setup of the connection to the leaf that requested to join. Once the call is active, however, managing of the connection to new and existing leaves is all done by the network.

In the case of *root LIJ*, the root adds and removes all leaves. When a leaf generates an LIJ request across its UNI to join a point-to-multipoint connection, the root is informed, which in turn extends the connection to the party.

The LIJ capability is not supported in the ATM Forum PNNI version 1.0. It is left as a work item for completion in PNNI version 2.0. Until this capability is defined in PNNI 2.0, there is no standard means to support LIJ in a network with switches from different vendors.

The LIJ capability requires a new state machine. The LIJ state machine, consisting of two states, exists only at the user side of the UNI where a leaf is attempting to add or drop itself to/from a connection:

- *Leaf-setup initiated*—Entered when a LEAF SETUP REQUEST message has been sent to the network side of the UNI interface;
- *Null*—Entered when a SETUP, ADD PARTY, or LEAF SETUP FAILURE message has been received on the user side of the interface, or when timer T331 expires.

3.10.1 Leaf-Initiated Join Messages

Two new messages are defined to support the LIJ capability: LEAF SETUP FAILURE and LEAF SETUP REQUEST. Unlike point-to-point and point-to-multipoint signaling messages, the LIJ messages are sent using a dummy call reference.

3.10.1.1 LEAF SETUP FAILURE

This message is sent to the leaf by the root or by the network to indicate that the request by a leaf to join a point-to-multipoint call is failed (see Figure 3.24).

The called party number IE is mandatory in the user-to-network direction. It is also mandatory in the network-to-user direction if the calling party number was included in the LEAF SETUP REQUEST message. Similarly, a called party subaddress is mandatory if the LEAF SETUP REQUEST message contained a calling party subaddress; otherwise, it is not allowed.

3.10.1.2 LEAF SETUP REQUEST

This message is sent to initiate leaf joining procedures (see Figure 3.25).

The called party number is the ATM address of the root. An LIJ call identifier is used to identify the call the leaf is attempting to join. This identifier by

Information Element	Type	Length
Cause	M	6–34
Called party number	O	≤25
Called party subaddress	O	4–25
Leaf sequence number	M	8
Transit network selection	O	4–8

Figure 3.24 Information elements used in LEAF SETUP FAILURE message.

Information Element	Type	Length
Transit network selection	O	4–8
Calling party number	M	≤26
Calling party subaddress	O	4–25
Called party number	M	≤25
Called party subaddress	O	4–25
LIJ call identifier	M	9
Leaf sequence number	M	8

Figure 3.25 Information elements used in LEAF SETUP REQUEST message.

itself is not unique in the network. In particular, the called party number and the LIJ call identifier together uniquely identify the point-to-multipoint call.

3.10.2 Leaf-Initiated Join Information Elements

LIJ-specific IEs are the *LIJ call identifier* IE, *LIJ parameters* IE, and *LIJ leaf sequence number* IE.

3.10.2.1 Leaf-Initiated Join Call Identifier IE

The LIJ call identifier IE is used to uniquely identify a point-to-multipoint call at a root's interface (see Figure 3.26). The LIJ call identifier is specified in the SETUP message when the root creates a point-to-multipoint call with LIJ capability.

This identifier is specified in the LEAF SETUP REQUEST message when a leaf wishes to join the identified call. It is noted that UNI 4.0 does not specify how a leaf knows an LIJ call identifier. The only LIJ call identifier type currently defined is *root assigned*, which indicates that the root specified the identifier value of the LIJ call identifier. The identifier value is a 4-octet value used to differentiate calls created by the same root. It is used by the network to distinguish one LIJ call from the other LIJ calls that this root creates. Hence, each LIJ call must have a unique LIJ call identifier. Inside the network, the LIJ call iden-

1	0 0 0	type	5
ext	spare		
LIJ call identifier value			6–9

Figure 3.26 LIJ call identifier IE.

tifier is paired with the root's address to form a common identifier that uniquely differentiates this network LIJ call from all others.

3.10.2.2 Leaf-Initiated Join Parameters IE

The LIJ parameters IE is used by the root to associate options with the call when the call is created (see Figure 3.27).

The only defined screening indicator value is "network join without root notification." Hence, this IE is only used if the root allows a leaf to join its call without root notification.

3.10.2.3 Leaf-Initiated Join Leaf Sequence Number IE

The LIJ leaf sequence number IE includes a 32-bit leaf sequence number field used by a joining leaf to associate a SETUP, ADD PARTY, or LEAF SETUP FAILURE response message with the corresponding LEAF SETUP REQUEST message that triggered the response (see Figure 3.28).

3.10.3 Root Creation of a Network Leaf-Initiated Join Call

In the case of a network LIJ call, the network automatically attempts to join requesting leaves. This type of a call is created by the root using the regular point-to-multipoint procedures at the UNI. One difference is that the SETUP message used to establish the connection to the first party contains additional IEs: LIJ leaf sequence number, LIJ parameters, and LIJ call identifier. All three IEs are mandatory to create a network LIJ call. Similarly, the ADD PARTY message is modified to include a leaf sequence number when a new party is added to a point-to-multipoint call in response to a LEAF SETUP REQUEST message.

If the LIJ parameters IE is specified without specifying both the leaf sequence number and the LIJ call IEs, the network rejects the setup request with cause "mandatory information element is missing." If the LIJ call parameters

1 ext	0 0 0 spare	Screening indicator	5

Figure 3.27 LIJ parameters IE.

Leaf sequence number	5–8

Figure 3.28 LIJ leaf sequence number IE.

information is not specified, the network assumes the creation of a root LIJ call (and does not automatically attempt to join leaves) (see Figure 3.29).

The remaining steps in the network LIJ call creation are identical to those of an ordinary point-to-multipoint call creation. After the call has been created, the root can freely add additional end points using the point-to-multipoint procedures. As with the first leaf, the SETUP messages to the new leaves may contain the new LIJ call identifier and LIJ parameters IEs.

3.10.4 Leaf Joins to Active Network Leaf-Initiated Join Call

When a leaf wishes to join a point-to-multipoint network LIJ call, it transfers a LEAF SETUP REQUEST message using the dummy call reference across the user-to-network interface and starts timer T331 (see Figure 3.30). The leaf is required to include a leaf sequence number IE in the LEAF SETUP REQUEST message to associate its LIJ call request with a particular call. This number should be unique (i.e., not in use). Using a unique identifier allows the leaf to later associate the SETUP, ADD PARTY, or LEAF SETUP FAILURE response message with the LIJ request.

To identify the call that a leaf wants to join, the LEAF SETUP RE-QUEST message must contain the LIJ call identifier IE and the root's ATM

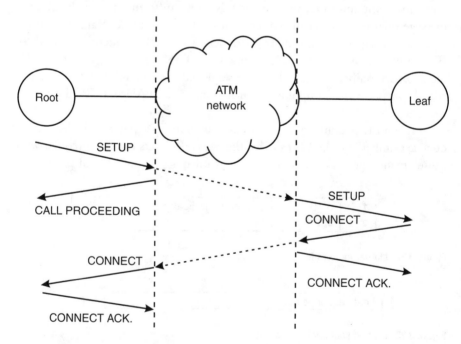

Figure 3.29 Root creation of a network LIJ call.

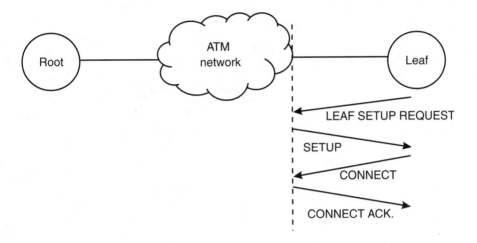

Figure 3.30 Leaf-initiated join to active network LIJ call.

address (in the called party number IE and, optionally, the called party sub-address IE). The leaf also includes its own address in the calling party number.

As an option, a leaf may include the transit network selection IE in the LEAF SETUP REQUEST message. If included, this IE is used by the network to route the LEAF SETUP REQUEST message towards the root using the specified transit network.

If timer T331 expires before receiving a response, the leaf may optionally retransmit the LEAF SETUP REQUEST message. The leaf may either use the same leaf sequence number as in the original request or select a new value. If it chooses not to retransmit the LEAF SETUP REQUEST message, or if timer T331 expires a second time, the leaf releases the leaf sequence number (see Figure 3.31).

Upon receiving the LEAF SETUP REQUEST message, the network does not change its state at the leaf's UNI. The called party number and called party subaddress (if specified) are used to internally forward the LEAF SETUP RE-QUEST message towards the root. If the network determines that the call information received from the user is invalid (e.g., invalid called party number), it sends a LEAF SETUP FAILURE message to the leaf using the dummy call reference and using the same leaf sequence number as that in the LEAF SETUP REQUEST message.

If the network can determine that access to the requested LIJ service is authorized and available, a SETUP or ADD PARTY message is sent to the leaf to add it to the requested LIJ call. The SETUP or ADD PARTY message sent to the leaf contains the leaf sequence number from the LEAF SETUP REQUEST message and may also contain the LIJ call identifier and LIJ parameters for this

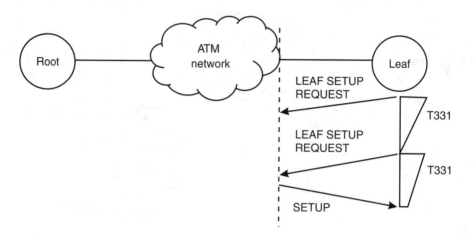

Figure 3.31 Timer T331 expiry.

call. If present in either the SETUP or ADD PARTY message, the calling party number and calling party subaddress will be that of the root, as specified in the corresponding IEs of the LEAF SETUP REQUEST message.

Upon receipt of the SETUP or ADD PARTY message, the leaf associates the leaf sequence number with the LIJ request and stops timer T331. The remaining steps are the same as for any other point-to-multipoint call setup.

The root does not get any notification of new leaves. It cannot determine how many leaves are receiving its transmissions and cannot selectively drop leaves that have added themselves. This feature, however, eliminates the possibility for the root to become a bottleneck for rapidly changing calls since it does not participate in the screening, notification, or drop procedures.

If the network is unable to complete the join request for any reason, a LEAF SETUP FAILURE message is returned to the leaf using the dummy call reference. The LEAF SETUP FAILURE message contains the following information from the LEAF SETUP REQUEST message: leaf sequence number, called party number (from the calling party number) and, optionally, the called party subaddress (from the calling party subaddress). Upon receiving the LEAF SETUP FAILURE message, timer T331 is stopped and the associated leaf sequence number is released.

3.10.5   Leaf Joins as the First Party to a Point-to-Multipoint Call

When a leaf issues a LEAF SETUP REQUEST message and it is the first party to join a point-to-multipoint call, the network uses the root's address and LIJ call identifier to form a common identifier. This identifier is used to identify and locate the indicated call. If the network cannot find the call (because this is the first party of the call), it forwards the LEAF SETUP REQUEST message to the root. Upon sending this message to the root, the network does not change its link state, does not enter a new party state, does not start any timers, or cache any internal information regarding the LEAF SETUP REQUEST message. The LEAF SETUP REQUEST message delivered to the root contains all of the IEs specified by the leaf.

Upon receiving the LEAF SETUP REQUEST message, the root has the option of creating a new call to the leaf by issuing a SETUP message or by sending a failure code back to the leaf in a LEAF SETUP FAILURE message.

Figure 3.32 shows the case in which the root chooses to create the call. The call establishment follows the existing point-to-multipoint procedures. The SETUP message must contain the leaf sequence number with a value equal to that given by the leaf in the LEAF SETUP REQUEST message. Optionally, the SETUP message may contain the LIJ call identifier and the LIJ parameters (if

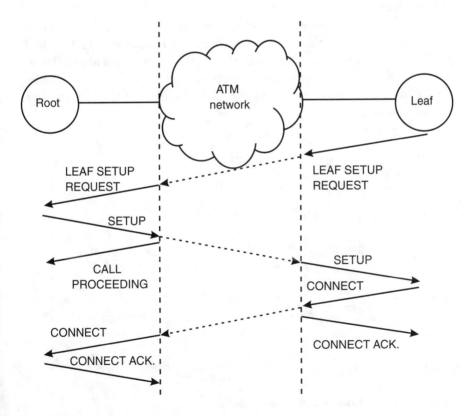

Figure 3.32 Leaf joined to inactive call.

the root chooses to create a network LIJ call, where automatic network joins are supported). The called party number value should be taken from the calling party number IE of the LEAF SETUP REQUEST message.

It is assumed that the LIJ call identifier is associated with the call parameters of the root. That is, the LIJ call identifier implies a certain traffic descriptor, quality of service, and so forth. If the root creates a call to the leaf with unacceptable call parameters, the leaf, upon receiving it, rejects the SETUP message.

If the root rejects the leaf's attempted join, it responds with a LEAF SETUP FAILURE message, as illustrated in Figure 3.33.

3.10.6   Leaf Joins to Root Leaf-Initiated Join Call

When a leaf attempts to join a root LIJ call, it sends a LEAF SETUP REQUEST message to the network and the network forwards the message to the root (see Figure 3.34). When the root receives this message, it either rejects the request by sending a LEAF SETUP FAILURE or adds the leaf by issuing an ADD PARTY message.

In the latter case, the existing point-to-multipoint procedures are followed with a minor exception that the leaf sequence number from the LEAF SETUP REQUEST message is included in the ADD PARTY message. The called party number and called party subaddress are taken from the calling party number and calling party subaddress IEs of the LEAF SETUP REQUEST message, respectively.

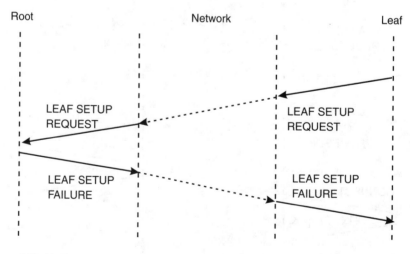

Figure 3.33 Unsuccessful leaf join to an inactive call.

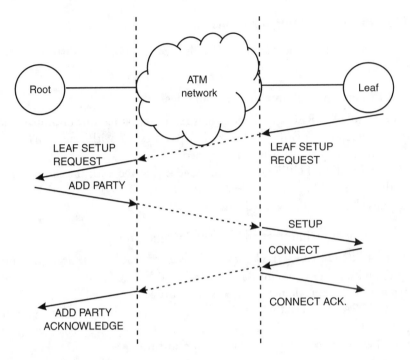

Figure 3.34 Leaf join to an active root LIJ call.

3.10.7   Call/Connection and Party Clearing

Call/connection and party clearing for the LIJ calls follow the corresponding point-to-multipoint clearing procedures with two exceptions.

If the root drops the last party it added in a network LIJ connection, it should do so by sending a DROP PARTY message rather than a RELEASE message, and continue to keep the call in the active link state. The network maintains the call in the active link state until the last leaf drops itself. In the meantime, the root can add additional parties. If the root wants to clear the call, it can send a RELEASE message at any time.

When the network drops the last party added by the root in a network LIJ connection while other leaves (that have added themselves) are still participating in the call, the network sends a DROP PARTY message rather than a RE-LEASE message and continues to keep the call in the active link state. Upon receipt of the DROP PARTY message, the root should likewise maintain the call in the active link state. The network sends a RELEASE message when the last leaf to add itself drops out of the call.

3.10.8   Handling of Error Conditions

The error-handling procedures of LIJ calls are the same as point-to-multipoint calls with the following additions.

3.10.8.1 Mandatory Information Element Missing

If a LEAF SETUP REQUEST message is received that has one or more mandatory IEs missing, a LEAF SETUP FAILURE message with cause "mandatory information element is missing" is returned.

If a LEAF SETUP FAILURE message is received with the cause IE missing, the actions taken are the same as if a LEAF SETUP FAILURE message with cause "normal, unspecified" was received.

3.10.8.2 Mandatory Information Element Content Error

If a LEAF SETUP REQUEST message is received that has one or more mandatory IEs with invalid contents, a LEAF SETUP FAILURE message with cause "invalid information element contents" is returned.

If a LEAF SETUP FAILURE message is received with invalid contents in the cause IE, it will be assumed that a LEAF SETUP FAILURE message was received with a missing cause information element.

3.10.9   Timers at the User Side

The timers used in LIJ procedures are listed in Table 3.5.

Table 3.5
The Timers Used in LIJ Procedures

Timer Number	Default Time-out Value	State of Call	Cause for Start	Normal Stop	At the First Expiry	At the Second Expiry	Implementation
T331	60 sec.	Leaf setup initiated LIJ state	LEAF SETUP REQUEST sent	SETUP, ADD PARTY, or LEAF SETUP FAILURE received	Optionally retransmit LEAF SETUP REQUEST and restart T331	Clear internal call, enter null LIJ state	Mandatory

4

Signaling Support for ATM Transfer Capabilities

When a connection request is accepted by the network, the user and the network agree upon a traffic contract for the duration of the connection. With this contract, the network guarantees the requested service demand of the connection as long as the source traffic stays within the specified limits. Accordingly, a traffic contract includes traffic descriptors, service requirements, and the conformance definition. Each one of these are required to be well defined and understood by both the network and the user.

ATM networks are expected to support a variety of services with different traffic characteristics. Design and operation of network control functions such as call admission, bandwidth reservation, and congestion control require accurate source characterization to achieve high resource utilization. However, a source, in general, cannot provide a detailed description of its traffic behavior. Hence, there is a trade-off between how much information should and can be defined to characterize a source.

A traffic parameter is a specification of a particular traffic aspect of the requested connection. The four parameters currently defined are the peak cell rate (PCR), cell delay variation tolerance (CDVT), sustainable cell rate (SCR), and burst tolerance (BT).

The PCR of a connection is the inverse of the minimum time between two cell submissions to the network. For example, if the PCR of a connection is 15 cells/s, cells may arrive at the switch at the rate of one cell in every two cell transmission times at the link when the link transmission rate is 30 cells/s. However, if the PCR of the connection is 20 cells/s, then two cells will arrive in every three cell slots. This would mean that some cells will arrive in consecutive slots. Based on the above definition, the PCR of the connection observed at the switch in this case would be 30 cells/s. The interarrival times of cells of a particular connection as monitored at the switch may also be affected when cells from two or more connections are multiplexed at the ATM layer. Furthermore, physical layer overhead and/or idle cells are inserted to the cell stream at the physical layer. All these events can potentially change the PCR of a connection as seen by the network. Cell delay variation tolerance is defined to provide a solution to these types of problems at an interface. Conceptually, CDVT defines how much deviation from the specified minimum interarrival time of cells is allowed at the UNI (i.e., as observed by the network). CDVT is specified by the network provider and it is not negotiated between the user and the network during the connection setup time.

Given the PCR and the CDVT, the maximum number of cells that can arrive back to back at the transmission link rate N is given as follows:

$$N = \lfloor 1 + CDVT / [(1 / PCR) - \delta] \rfloor \quad \text{for } 1/PCR > \delta$$

where $\lfloor x \rfloor$ denotes the integer part of x, and δ is the cell transmission time at the physical link.

The average rate of a connection is equal to the total number of cells transmitted divided by the duration of the connection. Based on this definition, the network can know the average rate of a connection only after the connection terminates. Sustainable cell rate is an upper bound on the average rate of an ATM connection that can be monitored by the network throughout the duration of the connection. SCR is used together with another metric, the BT and the PCR. BT is the duration of the period the source is allowed to submit traffic at its peak cell rate. Given these parameters, the maximum number of cells that can arrive at the switch back to back (i.e., at the peak cell rate), referred to as the maximum burst size (MBS), is given as follows:

$$MBS = \lfloor 1 + BT / [(1 / SCR) - (1 / PCR)] \rfloor$$

The cell loss priority (CLP) bit at the ATM cell header defines two loss priorities: high priority (CLP = 0) and low priority (CLP = 1). This bit may be used mainly for two purposes. An application may want to define to the network which cells of its traffic are more important than the others. Another use of high/low loss priority at the cell header is referred to as *tagging*. The network polices arrival cell streams based on the traffic contracts of connections to determine whether or not connections stay within the source traffic parameters agreed upon at the call-establishment phase. When a cell is detected to be non-conforming, there are two choices: either drop the cell at the interface or allow the cell to enter the network, hoping that there might be enough resources to deliver the cell to its destination. In the latter case, it is necessary to make sure that nonconforming cells do not cause degradation to the service provided to connections that stay within their negotiated parameters. Tagging is allowing nonconforming cells with CLP = 0 to enter the network with CLP = 1 (i.e., the CLP bit is changed from 0 to 1). Cells with CLP = 1 are discarded in the network whenever necessary (i.e., when there is a congestion) so that the service provided to the conforming traffic (i.e., in this context, CLP = 0 traffic) is not affected.

In summary, the combination of traffic parameters allowed in signaling are as shown in Table 4.1.

Table 4.1
Traffic Parameters Used in Signaling

Combination	Traffic Parameters
1	PCR(0); PCR(0 + 1); tagging is not requested
2	PCR(0); PCR(0 + 1); tagging is requested
3	PCR(0 + 1); SCR(0); BT; tagging is not requested
4	PCR(0 + 1); SCR(0); BT; tagging is requested
5	PCR(0 + 1)
6	PCR(0 + 1); SCR(0 + 1); BT; tagging is not requested

Users at the setup phase request an ATM layer quality of service selected from the QoS classes the network provides for ATM layer connections. A QoS class has either specified performance parameters (referred to as specified QoS class) or no specified performance parameter (referred to as unspecified QoS class). The former provides a quality of service to an ATM connection in terms of a subset of performance parameters (defined next). In the latter case, there is no explicitly specified QoS commitment on the cell flow. This class is intended to be used for best-effort service.

Currently supported QoS classes at the UNI are as follows:

- *unspecified QoS class 0*—supports best-effort service in which no explicit service requirement is specified;
- *specified QoS class 1*—supports constant bit rate traffic with stringent timing requirements, such as circuit emulation;
- *specified QoS class 2*—supports variable bit rate traffic with stringent timing requirements, such as compressed voice and video;
- *specified QoS class 3*—supports connection-oriented, variable bit rate traffic with no timing requirements, such as frame relay;
- *specified QoS class 4*—supports connectionless, variable bit rate traffic with no timing requirements, such as IP over ATM.

Different B-ISDN applications have different performance requirements. Accordingly, B-ISDNs are expected to support a variety of performance levels within a particular QoS class. To support this requirement, ATM signaling allows the individual performance metrics to be exchanged between the network and the end stations. In general, the QoS requirement of an application is defined through a number of parameters that may include one or more of the parameters listed in Table 4.2.

Table 4.2
Some QoS Parameters Used in ATM

Parameter	Description
Cell error ratio (CER)	The ratio of the number or cells with errors in the cell payload to the total number of cells transmitted by the source end station
Cell misinsertion rate (CMR)	The ratio of cells delivered to a wrong destination to the total number of cells sent
Cell loss ratio (CLR)	The ratio of the number of cells lost to the total number of cells transmitted by the source over the lifetime of the connection
Maximum cell transfer delay (CTD)	The end-to-end delay is the sum of various fixed (i.e., propagation delay) and random delay components (i.e., queuing delay). The maximum peak-to-peak CTD specified for a connection is the $(1 - 10^{-\alpha})$ quantile of the CTD.
Cell delay variation (CDV)	$(1 - 10^{-\alpha})$ quantile of CTD minus fixed CTD that could be experienced by any delivered cell on a connection during the entire connection holding time

Table 4.3 illustrates some of the values of the QoS parameters defined in I.356 for different QoS classes.

Table 4.3
QoS Classes Defined in Signaling

QoS Class	QoS Parameter					
	CTD	CDV	CLR(0+1)	CLR(0)	CER	CMR
QoS class 1	400 ms	3 ms	3×10^{-7}	None	None	None
QoS class 2	Unspecified	Unspecified	1×10^{-5}	None	4×10^{-6}	1 per day
QoS class 3	Unspecified	Unspecified	Unspecified	1×10^{-5}	4×10^{-6}	1 per day

4.1 ATM Transfer Capabilities

An ATM transfer capability supports ATM layer services provided at the ATM layer. These service categories relate the traffic characteristics and QoS requirements of a connection to a network behavior. The user's view of an ATC is the service suitable for a particular set of applications. From a network operator's perspective, an ATC may provide higher resource utilization through statistical multiplexing.

It is a network option to support one or more of the specified ATCs. It is mandatory that the ATM transfer capability used on a given ATM connection be implicitly or explicitly declared during the connection-establishment time.

There is no one-to-one correspondence between the QoS service classes and ATM transfer capabilities. As a result, the requested ATM transfer capability (as supported by signaling) should not be checked by call admission control (CAC) against any other information than what is contained in the traffic contract (i.e., QoS class, source traffic descriptor, and associated CDV tolerances). VCCs with the same ATC may be multiplexed into a VPC with the same ATC. In addition, a VCC or VPC with a particular ATC may be used to emulate a VCC or VPC with any different ATC.

ATM Forum UNI 4.0 supports the following ATCs:

1. Constant bit rate (CBR) [or, deterministic bit rate (DBR)];
2. Real-time variable bit rate (RT-VBR) [or, statistical bit rate (SBR)];

3. Non-real-time variable bit rate (NRT-VBR) [or, statistical bit rate (SBR)];

4. Available bit rate (ABR);

5. Unspecified bit rate (UBR).

CBR service class is used for real-time applications that require tightly constrained delay and delay variation, such as voice and video. A consistent availability of a fixed quantity of bandwidth equal to PCR (or more, depending on the CTDV specification, internal switch architecture, and so forth) is required in the network for CBR applications.

The RT-VBR service class, similar to the CBR service class, is intended for real-time applications that require tightly constrained delay and delay variation, such as voice and video. RT-VBR sources generate traffic at a rate that varies over time (i.e., bursty sources). This service allows statistical multiplexing and consistent service quality to applications such as compressed voice and video. In this context, when two or more connections are statistically multiplexed onto a link, the sum of their peak rates exceeds the link transmission rate while the sum of their average cell rates is less than the link transmission rate. For example, when 10 connections each with PCR = 10 Mbps and average cell rate of 1 Mbps are multiplexed onto a 25-Mbps link, the sum of the peak rates is equal to 100 (much larger than the link transmission rate), whereas the average utilization at the link is equal to 0.4 (i.e., 10/25). If statistical multiplexing is not used, then only two such connections can be multiplexed onto the link (i.e., so that 20 Mbps < 25 Mbps), which means the average utilization of the link is 0.08.

The traffic characteristics of applications that use RT-VBR service are characterized by their PCR, CTDV, SCR, and BT. The amount of bandwidth required to support this type of traffic in the network is determined by various factors such as buffer size, scheduling discipline, and CLR requirements. The required bandwidth is a value between the SCR and the PCR. The specific bandwidth required to support a VBR source is determined in a vendor-specific manner and is not a part of any standards activity.

The NRT-VBR service class is defined for non-real-time applications with bursty traffic characteristics. The traffic characteristics of applications that use NRT-VBR service are characterized by their PCR, CTDV, SCR, and BT. Similar to the RT-VBR service, the amount of bandwidth required to support a non-real time is vendor specific and is not a part of any standards activity.

The UBR service class supports delay-tolerant applications that are not sensitive to delay variation such as traditional computer communications applications. Unlike the other three service classes discussed so far, there is no bandwidth reservation required to support UBR service in the network (i.e., provides best-effort service). Accordingly, there is no service guarantee provided at the ATM layer. The end-to-end data integrity may be achieved by the higher layer protocols. The traffic characteristics of applications that use the UBR service are characterized by their PCR and CTDV.

The UBR service is intended to support connectionless data traffic that does not require QoS guarantees from the network. This type of traffic increases the utilization of network resources beyond what can be achieved by reserved traffic alone (i.e., CBR, real-time, and NRT-VBR services). Conceptually, cells belonging to the UBR service are transmitted only when there is no cell waiting to be transmitted in the transmission link buffer. If service guarantees are required, NRT-VBR service may be used instead. However, VBR service requires the specification of SCR and BT, which may not be known a priori and/or the cost associated with VBR service may not be justified (or preferred) by some applications (i.e., electronic mail). UBR service, however, may not be the best way to support applications that have the capability to adapt to network congestion status. In particular, when cells of UBR applications are lost in the network, this information will be known only through the use of higher layer protocols at the end stations. The consequence of this might be that most of the UBR traffic carried might be retransmissions, causing significant throughput degradation and eventually reaching a catastrophic state in which all cells carried by the UBR service are retransmissions.

Various applications [e.g., Transmission Control Protocol (TCP)] have the ability to react to congestion in the network by adjusting their information transfer rates based on feedback from the network or peer application. Such sources may increase their rate if there is extra bandwidth available and scale it down if there is a lack of it.

Under these conditions, there is a need for the network to inform the application of impending congestion via a feedback mechanism to the sender so that the source can vary its information transmission rate up and down between a specified lower (i.e., zero) and an upper (i.e., peak cell rate) bound, based on this feedback. ABR service is defined to support this type of traffic.

In the ABR service, cell transfer rate is adjusted at the end stations based on the information provided by the network. The traffic parameters specified are the PCR and CDVT, which is used to define the upper bound on the information transfer rate and the minimum cell rate (MCR) that specifies the minimum usable bandwidth. MCR could be any value between 0 and PCR, depending on the application requirements. The network (implicitly) guarantees to provide a low CLR to sources that adjust their cell flows according to the feedback information.

4.1.1 Signaling of ATM Transfer Capabilities

The broadband bearer capability and ATM traffic descriptor IEs are changed to support different ATCs.

4.1.1.1 Broadband Bearer Capability

Signaling of ATC is specified in Q.2961 (part 2). In particular, octet 5a of the broadband bearer capability IE was restructured to support the ATCs. This restructuring is backward-compatible with the first edition of Q.2931, which supports only a subset of the following ATCs. The format of the revised broadband bearer capability IE is shown in Figure 4.1.

Five bearer classes are defined (octet 5) as listed in Figure 4.2.

BCOB-A (broadband connection oriented bearer, class A) is a connection-oriented service that provides a constant bit rate service. It is defined for use by applications with stringent timing requirements such as voice. BCOB-A is used only for a virtual channel service. When BCOB-A is specified, the user may be requesting more than an ATM-only service. The network may look at the AAL IE to provide interworking based upon its contents. An example would be interworking with a N-ISDN circuit switching service (cf. Chapter 6).

BCOB-C (broadband connection oriented bearer, class C) is also a connection-oriented service. Applications that use this service generate variable bit

0/1 ext	spare	Bearer Class		5
1 ext	Broadband transfer capability (BTC)			5a*
1 ext	Susceptibilty to clipping	Spare	User plane connection configuration	6

Figure 4.1 Broadband bearer capability IE revised to support ATCs.

Field	Bearer Class
0 0 0 0 1	BCOB-A
0 0 0 1 1	BCOB-C
0 0 1 0 1	Frame Relay bearer service
1 0 0 0 0	BCOB-X
1 1 0 0 0	Transparent VP Service

CBR (only f / v VBR —, defined by stu wo [handwritten annotation]

Figure 4.2 Bearer class field.

rate traffic. It supports applications with timing requirements as well as others that do not impose any timing restrictions. BCOB-C is used for a virtual channel service only. As for BCOB-A, when a BCOB-C ATC is specified, the user may be requesting more than an ATM-only service and the network may look at the AAL IE to provide interworking based upon its contents.

BCOB-X is a connection-oriented service in which the timing requirements and traffic type are defined by the user. BCOB-X is used for requesting a virtual channel service. This service is used to request an ATM-only service from the network and the network does not process any higher layer protocols (e.g., AAL). A VBR user that wants only an ATM cell relay service would use BCOB-X. Similarly, a user that is placing a DS1 circuit emulation call but does not want to allow interworking would use BCOB-X as well. If the user wishes to allow interworking, then the BCOB-A service is specified.

The user may also request a transparent VP service. It differs from BCOB-X in that with the transparent VP service, both the VCI field (except for VCI values 0, 3, 4, and 6 through 15) and payload type field are transported transparently by the network.

The values of the broadband transfer capability (BTC) field are defined as shown in Figure 4.3.

Constant bit rate with end-to-end timing required, variable bit rate with end-to-end timing required, and variable bit rate with end-to-end timing not required code points are provided for backward compatibility with earlier versions of Q.2931. The details of each BTC are given in subsequent sections.

4.1.1.2 ATM Traffic Descriptor

The ATM traffic descriptor IE is used to specify the set of traffic parameters (see Figure 4.4). Except for the forward and backward PCR for CLP = 0 + 1, all other parameters are optional and position-independent. If PCR for

Field	Description
0 0 0 0 1 0 1	Constant bit rate (CBR) with end-to-end timing required
0 0 0 0 1 1 1	Deterministic bit rate (DBR)
0 0 0 1 0 0 1	Variable bit rate (VBR) with end-to-end timing required
0 0 0 1 0 1 0	Variable bit rate with end-to-end timing not required
0 0 0 1 0 1 1	Statistical bit rate (SBR) with end-to-end timing not required
0 0 1 0 0 1 1	Statistical bit rate with end-to-end timing required
0 0 1 0 0 0 0	ATM block transfer (ABT)-delayed transmission
0 0 1 0 0 0 1	ATM block transfer-immediate transmission
0 0 0 1 1 0 0	Available bit rate

Figure 4.3 Broadband transfer capability field.

Field	Octet
Forward Peak Cell Rate identifier (for CLP = 0)	5*
Forward Peak Cell Rate (for CLP = 0)	5.1 to 5.3*
Backward Peak Cell Rate identifier (CLP = 0)	6*
Backward Peak Cell Rate (for CLP = 0)	6.1 to 6.3*
Forward Peak Cell Rate identifier (for CLP = 0 + 1)	7
Forward Peak Cell Rate (for CLP = 0 + 1)	7.1 to 7.3
Backward Peak Cell Rate identifier (for CLP = 0 + 1)	8
Backward Peak Cell Rate (for CLP = 0 + 1)	8.1 to 8.3
Forward Sustainable Cell Rate identifier (for CLP = 0)	9*
Forward Sustainable Cell Rate (for CLP = 0)	9.1 to 9.3*
Backward Sustainable Cell Rate identifier (for CLP = 0)	10*
Backward Sustainable Cell Rate (for CLP = 0)	10.1 to 10.3*
Forward Sustainable Cell Rate identifier (for CLP = 0 + 1)	11*
Forward Sustainable Cell Rate (for CLP = 0 + 1)	11.1 to 11.3*
Backward Sustainable Cell Rate identifier (for CLP = 0 + 1)	12*
Backward Sustainable Cell Rate (for CLP = 0 + 1)	12.1 to 12.3*
Forward Maximum Burst Size identifier (for CLP = 0)	13*
Forward Maximum Burst Size (for CLP = 0)	13.1 to 13.3*
Backward Maximum Burst Size identifier (for CLP = 0)	14*
Backward Maximum Burst Size (for CLP = 0)	14.1 to 14.3*
Forward Maximum Burst Size identifier (for CLP = 0 + 1)	15*
Forward Maximum Burst Size (for CLP = 0 + 1)	15.1 to 15.3*
Backward Maximum Burst Size identifier (for CLP = 0 + 1)	16*
Backward Maximum Burst Size (for CLP = 0 + 1)	16.1 to 16.3*
Best Effort Indicator	17*
Traffic Management Options Identifier	18*

Forward frame discard	Backward frame discard	0	0	0	0	Tagging backward	Tagging forward	18.1*

Figure 4.4 ATM traffic descriptor IE.

CLP = 0 + 1 is the only parameter specified, the network resource allocation assumes the entire peak cell rate can be used for CLP = 0.

The valid combinations of the traffic descriptor subfields in the ATM traffic descriptor IE for different ATM transfer capabilities are discussed in the subsequent sections.

4.2 Deterministic Bit Rate Transfer Capability

The DBR transfer capability is used by connections that request a static amount of bandwidth that is continuously available during the connection lifetime. The amount of bandwidth is characterized by the connection's peak cell rate value.

The basic commitment made by the network to a user who reserves resources via the DBR capability is that once the connection request is accepted by the network, the negotiated ATM layer QoS is assured to all cells when all cells are conforming to the relevant conformance tests. That is, the source, in the DBR capability, can emit cells at the peak cell rate at any time and for any duration and its QoS requirements are met in the network.

The DBR transfer capability is intended to support real-time applications requiring tightly constrained delay variation (e.g., voice, circuit emulation), but it is not restricted to these applications. In the DBR capability, the source may emit cells at or below the negotiated PCR (and may also even be silent) for periods of time.

The DBR capability may be used for both VPCs and VCCs. The tagging option or selective cell discard of CLP = 1 cells does not apply to connections with the DBR capability. Table 4.4 defines allowable combinations of QoS and traffic parameters used in the DBR service.

Table 4.4
Allowable Combinations of Traffic and QoS-Related Parameters for DBR in the SETUP Message

Conformance	Used When the CLR Commitment on CLP = 0 + 1 Traffic Is Required			Included for Backward Compatibility With UNI 3.1 and ITU-T Recommendation I.371		Included for Backward Compatibility With ATM Forum UNI 3.1 and ITU-T Recommendation I.371	
Bearer Capability							
Broadband Bearer Class	A	X	VP	A	X	A	X
Broadband Transfer Capability	DBR			Absent	CBR	Absent	CBR
Traffic Descriptor for a Given Direction							
PCR (CLP = 0)	Unspecified			Unspecified		Specified	
PCR (CLP = 0 + 1)	Specified			Specified		Specified	
Tagging	No			No		Yes/No	
Frame discard	Yes/No			Yes/No		Yes/No	

Table 4.4 (Continued)

Conformance	Used When the CLR Commitment on CLP = 0 + 1 Traffic Is Required	Included for Backward Compatibility With UNI 3.1 and ITU-T Recommendation I.371	Included for Backward Compatibility With ATM Forum UNI 3.1 and ITU-T Recommendation I.371
QoS classes	Network option	Network option	Network option
Transit delay and extended QoS parameters	Mandatory when the QoS parameter IE (QoS classes) is absent. Extended QoS parameters are the CDV and CLR.		

4.3 Statistical Bit Rate Transfer Capability

The SBR transfer capability is defined for applications that have bursty traffic characteristics. In addition to PCR, connections with the SBR capability use the SCR and MBS to characterize their traffic in greater detail. This allows the network to use its resources more efficiently through statistical multiplexing when such a connection is multiplexed together with other connections in the network.

The SBR capability is suitable for applications when an estimate of SCR and MBS are known a priori. SBR capability can be used for real-time applications (i.e., those requiring tightly constrained delay and delay variation). In this case, cells that are delayed beyond the value specified by the maximum CTD (maxCTD) are assumed to be of significantly reduced value to the application. The SBR transfer capability can also be used for non-real-time applications (i.e., applications with bursty traffic characteristics that do not require any strict delay bounds). However, these applications may require a low cell loss ratio.

The SBR capability may be used for both VPCs and VCCs. The tagging option may be supported and the selective cell discard function may be applied to these connections.

4.3.1 Signaling Procedures for the SBR Transfer Capability

The calling party initiates call establishment using the Q.2931 procedures. The rules for selecting the traffic parameters in a given direction are specified as follows:

- PCR for CLP = (0 + 1) is a mandatory parameter.
- The value of the sustainable cell rate must be less than the PCR.
- The tagging option may only be used when the ATM traffic descriptor IE includes a parameter on CLP = 0.
- The SCR and MBS must be provided together for the same CLP indication.

The traffic parameters for the forward and backward directions are specified independent of each other (i.e., the forward direction may use one combination of traffic parameters, while the backward direction uses another combination of traffic parameters). Tables 4.5 and 4.6 define allowable combinations of QoS and traffic parameters used in RT-SBR service and NRT-SBR service, respectively.

4.4 Available Bit Rate Transfer Capability

ABR is an ATM transfer capability that allows an end station to regulate its traffic rate based on a feedback received from the network and may change subsequent to the connection establishment. A user that adapts its traffic to the changing ATM layer transfer characteristics is expected to experience a low cell loss ratio. The two other QoS parameters, cell delay variation and cell transfer delay, are not controlled. This service is designed for applications that have the ability to reduce their information transfer rate if the network requires them to do so. Likewise, they may wish to increase their information transfer rate if there is extra bandwidth available within the network.

The user adapts to the changing ATM layer transfer characteristics upon receiving a feedback from the network. Due to the cell transfer delay, this feedback reflects the status of the network at some time prior to the instant the user receives it. So, even when the user adapts correctly to the feedback, the network may still have to provide enough buffering to enable low cell loss operation.

During the call establishment time, the calling user specifies its maximum required bandwidth (peak cell rate) to the network. The PCR of an ABR connection is negotiated between the user and the network and between two end stations during the connection-establishment time. A minimum useable bandwidth [referred to as the minimum cell rate (MCR)] is also specified on a per-connection basis. MCR can be specified as zero. A source is not precluded from sending at a rate less than the MCR if the MCR is negotiated to be greater than zero. The value of the PCR and the MCR can be different in each direction of a bidirectional connection.

Table 4.5
Allowable Combinations of Traffic and QoS-Related Parameters for RT-SBR in the
SETUP Message

Conformance	SBR.1 (Used When the CLR Commitment on CLP = 0 + 1 Traffic Is Required)			SBR.2 (Used When the CLR Commitment on CLP = 0 Traffic Is Required)			SBR.3 (Used When the CLR Commitment on CLP = 0 Traffic Is Required)		
	Bearer Capability								
Broadband Bearer Class	C	X	VP	C	X	VP	C	X	VP
Broadband Transfer Capability	SBR with end-to-end timing requirements			VBR with end-to-end timing requirements			VBR with end-to-end timing requirements		
	Traffic Descriptor for a Given Direction								
PCR (CLP = 0)	Unspecified			Unspecified			Unspecified		
PCR (CLP = 0 + 1)	Specified			Specified			Specified		
SCR, MBS (CLP = 0)	Unspecified			Specified			Specified		
SCR, MBS (CLP = 0 + 1)	Specified			Unspecified			Unspecified		
Tagging	No			No			Yes		
Frame discard	Yes/No			Yes/No			Yes/No		
QoS classes	Network option			Network option			Network option		
Transit delay and extended QoS parameters	Mandatory when the QoS parameter IE (QoS classes) is absent. CDV and CLR are the two extended QoS parameters.								

An ABR service is currently applicable only to VCCs, and it is not yet defined for the VPCs. In ABR, the CLP bit of the user data cell is set to 0 and tagging is not supported.

ABR flow control occurs between a sending end station (source) and a receiving end station (destination). The source sends a resource management (RM) cell periodically. Switches in the network and the destination end station manipulate these RM cells to provide feedback to the source. For a bidirectional ABR connection, each of the end stations is both a source and a destination.

Table 4.6
Allowable Combinations of Traffic and QoS-Related Parameters for NRT-SBR in the SETUP Message

Conformance	SBR.1 (Used When the CLR Commitment on CLP = 0 + 1 Traffic Is Required)			SBR.2 (Used When the CLR Commitment on CLP = 0 Traffic Is Required)			SBR.3 (Used When the CLR Commitment on CLP = 0 Traffic Is Required)		
Bearer Capability									
Broadband Bearer Class	C	X	VP	C	X	VP	C	X	VP
Broadband Transfer Capability	SBR with end-to-end timing is not required			Absent	VBR with no timing requirement	VBR with no timing requirement	Absent	VBR with no timing requirement	VBR with no timing requirement
Traffic Descriptor for a Given Direction									
PCR (CLP = 0)	Unspecified			Unspecified			Unspecified		
PCR (CLP = 0 + 1)	Specified			Specified			Specified		
SCR, MBS (CLP = 0)	Unspecified			Specified			Specified		
SCR, MBS (CLP = 0 + 1)	Specified			Unspecified			Unspecified		
Tagging	No			No			Yes		
Frame discard	Yes/No			Yes/No			Yes/No		
QoS classes	Network option			Network option			Network option		
CLR	Mandatory when the QoS parameter IE (QoS classes) is absent								
Max. transit delay in forward direction	May be specified for the ATM service category of NRT-SBR for reasons of backward compatibility with ITU-T Recommendation Q.2931								

There are three ways a switch (or the destination end station) can provide a feedback to the source end stations. First, each cell header contains an explicit forward congestion indication (EFCI) that can be set by a congested switch. Second, RM cells have two bits in their payload: congestion indicator (CI) and no increase (NI). These two bits can be set by the switch based on its internal status. Finally, RM cells have another field in the payload, referred to as the explicit rate (ER). A switch can set the value of this field to the minimum of the ER value received from the neighbor switch and the rate it can support for the particular connection to which the RM cell corresponds.

4.4.1 Signaling for ABR

An ABR connection is established using the Q.2931 messages and procedures. Additional traffic parameters are added to various Q.2931 messages to support this transfer capability.

4.4.1.1 Messages
A new information element, *ABR setup parameters* IE, is added to both CONNECT and SETUP messages. This IE is mandatory for the ABR transfer capability. In addition, the ATM traffic descriptor IE included in the CONNECT message contains the PCR and the ABR minimum cell rate subfields. (See Figures 4.5 and 4.6.)

The *minimum acceptable ATM traffic descriptor* IE is included in the SETUP message when the ABR MCR traffic parameter is negotiable.

4.4.1.2 Information Elements
Various information elements defined/modified to support the ABR service are defined in this section.

ATM Traffic Descriptor
The *ATM traffic descriptor* IE is extended to include ABR MCR, as shown in Figure 4.7.

Information Element	Length
ABR setup parameters	4–36
ATM traffic descriptor	12–20

Figure 4.5 ABR-related IEs used in the CONNECT message.

Information Element	Length
ABR setup parameters	4–36
ATM traffic descriptor	12–20
Minimum acceptable ATM traffic descriptor	4–12

Figure 4.6 ABR-related IEs used in the SETUP message.

Field	Octets
Forward ABR minimum cell rate identifier (CLP = 0 + 1)	19*
Forward ABR minimum cell rate (for CLP = 0 + 1)	19.1 to 19.3*
Backward ABR minimum cell rate identifier (CLP = 0 + 1)	20*
Backward ABR minimum cell rate (for CLP = 0 + 1)	20.1 to 20.3*

Figure 4.7 ATM traffic descriptor IE.

ABR Setup Parameters

The purpose of the ABR setup parameters IE is to specify the set of ABR parameters during call/connection establishment (see Figure 4.8). The maximum length of this IE is 36 octets.

All the parameters are position independent.

Except for the *cumulative RM fixed round-trip time* parameter, all other parameters are optional in the user-to-network direction in a SETUP message. However, all parameters are mandatory in the network-to-user direction in a SETUP message and in both directions in a CONNECT message.

The parameters *rate increase factor* (RIF) and *rate decrease factor* (RDF) are signaled as $\log_2$ (RIF * 32768) and $\log_2$ (RDF * 32768), respectively. The range of this parameter is 0–15. It is coded as an 8-bit binary integer, with bit 8 being the most significant bit and bit 1 being the least significant bit.

4.4.2 Signaling Procedures

The procedures for basic call/connection control procedures are defined in Q.2931. Only the additional procedures defined to handle the point-to-point ABR call/connections are described next.

4.4.2.1 Call/Connection Establishment at the Originating Interface

The calling party initiates ABR call establishment by sending a SETUP message across its UNI. The SETUP message contains a broadband bearer capability

Field	Octets
Forward ABR initial cell rate identifier (CLP=0+1)	5*
Forward ABR initial cell rate (for CLP=0+1)	5.1 to 5.3*
Backward ABR initial cell rate identifier (CLP=0+1)	6*
Backward ABR initial cell rate (for CLP=0+1)	6.1 to 6.3*
Forward transient buffer exposure identifier	7*
Forward transient buffer exposure	7.1 to 7.3*
Backward transient buffer exposure identifier	8*
Backward transient buffer exposure	8.1 to 8.3*
Cumulative RM fixed round-trip time identifier	9*
Cumulative RM fixed round-trip time	9.1 to 9.3*
Forward rate increment factor identifier	10*
Forward rate increment factor	10.1*
Backward rate increment factor identifier	11*
Backward rate increment factor	11.1*
Forward rate decrease factor identifier	12*
Forward rate decrease factor	12.1*
Backward rate decrease factor identifier	13*
Backward rate decrease factor	13.1*

Figure 4.8 ABR setup parameters.

IE indicating ABR service and point-to-point in the user plane connection configuration. The ABR setup parameters IE is also mandatory in the SETUP message.

ABR parameters for a given (forward or backward) direction may be included in the ABR setup parameters IE only if the ATM traffic descriptor IE contains a nonzero PCR (CLP = 0 + 1) value in that direction.

The following additional rules apply to the ABR ATM transfer capability:

- Tagging cannot be requested.
- The calling user may include the MCR parameter for one or both directions [provided that the PCR (CLP = 0 + 1) parameter value is different from zero in that direction].
- The cumulative RM fixed round-trip time parameter in the ABR setup parameters IE is set to the calling user's RM cell delay contribution for the forward and backward directions of the connection.
- In the ABR setup parameters IE, the calling user may include value(s) for ICR, TBE, RIF, and RDF in one or both directions.

- If the calling user wants to allow negotiation of the MCR parameter, the corresponding MCR parameter is included in the minimum acceptable ATM traffic descriptor parameters IE.

Various ABR parameters are assigned default values. If the user does not specify a value for a particular parameter, then its default value is assumed. Table 4.7 illustrates the ABR parameters and their associated default values.

MCR can be negotiated, using the procedures described in Q.2962, if the corresponding MCR parameter is included in the minimum acceptable ATM traffic descriptor IE in the SETUP message. If able to provide the indicated PCR and ABR setup parameter values, the network progresses the call towards the called user without changing the original parameters.

Table 4.7
ABR Parameters

Label	Parameter	Default Value
PCR	Peak cell rate	—
MCR	Minimum cell rate	0
ACR	Allowed cell rate	—
ICR	Initial cell rate	PCR
TCR	Tagged cell rate	10 cells/s
Nrm	Number of cells between forward RM cells	32
Mrm	Controls bandwidth allocation between forward RM cells, backward RM cells, and data cells	2
Trm	Upper bound on interforward RM cell time	100 ms
RIF	Rate increase factor	1/16
RDF	Rate decrease factor	1/16
ADTF	ACR decrease time factor	0.5 sec
TBE	Transient buffer exposure	16,777,215
CRM	Missing RM cell count	TBE/Nrm
CDF	Cutoff decrease factor	1/16
FRTT	Fixed round-trip time	—

If the indicated PCR cannot be supported, but at least the indicated MCR can be provided, the network progresses the call towards the called user, after adjusting the PCR parameter to the value that can be provided. The adjusted PCR value is always greater than or equal to the indicated MCR in the same direction.

ACR is the rate at which a source is allowed to send at any particular instant of time. It changes dynamically between MCR and PCR. After a period of time during which the source is inactive and when the connection becomes active, ACR is set to ICR.

RM cells are, in general, considered as part of the user traffic and are referred to as in-rate RM cells, reflecting the fact that the total rate of data and in-rate RM cells is less than or equal to ACR. Under exceptional circumstances, any ATM device along the end end-to-end path (including the end stations) can generate out-of-rate RM cells to react quickly to changing conditions. These cells are tagged (i.e., CLP = 1), and the network carries them only if it is not congested. The tagged cell rate (for out-of-rate RM cells) is set to 10 cells/s.

A source is required to send a forward RM cell (FRM) after every Nrm − 1 data cells. If the source rate is low, however, the time between RM cells may be too large for network feedback to be effective. To overcome this problem, the parameter Trm is defined. If more than Trm seconds have passed since the last time a source sent an FRM, an FRM cell is transmitted by the source, independent of the number of data cells transmitted since then. However, there must be at least Mrm − 1 data cells between two FRMs.

If a source does not send an RM cell for a period of time equal to ADTF, it may not be able to use its current ACR. In particular, if ACR is greater than ICR, a source is required to reduce its ACR to ICR if ADTF expires.

If one or more resources along its end-to-end path are broken or heavily congested, a source may not receive backward RM cells for the forward RM cells it transmitted. The CRM and CDF parameters are used to address this problem. In particular, if a source has transmitted CRM forward RM cells and has not received any backward RM cells, it is required to reduce its rate by a factor of CDF. One major problem with this solution is that if CRM is not set properly, a long delay path may artificially cause this rule to be triggered. To address this problem, CRM is calculated using another parameter. TBE is the maximum number of cells that may suddenly appear at a switch during a period equal to the first round trip, before an RM cell can be turned back to provide a feedback. During this time, a source would have sent TBE/Nrm cells. Hence, CRM is equal to TBE/Nrm.

FRTT is the minimum delay along the end-to-end path (excluding the queuing delays). During this time, a source may transmit ICR × FRTT cells into the network. TBE is greater than or equal to ICR × FRTT.

In a network with some switches manipulating only the EFCI bit and others providing ER, it is possible that a new ER returned in an RM cell could be much higher than the ACR. This might cause severe degradation to the cell loss performance at the EFCI switches. RIF is used to restrict the amount of increase allowed. In particular, the source is not allowed to increase its current ACR by more than RIF × PCR, even if the ER returned in the RM cell is greater than this value.

The CI bit in the RM cell payload is used to indicate congestion (either by switches or the destination end station). When CI is set to 1, the source is required to decrease its ACR by a factor of (1 − RDF); that is, the new ACR is equal to ACR × (1 − RDF). If this value is less than the MCR, then the ACR is set to MCR.

The NI bit is used to handle mild congestion periods. If NI is set to 1, the source is not allowed to increase its ACR. The actions corresponding to different values of NI and CI are shown in Table 4.8.

Table 4.8
The Use of NI and CI Bits

NI	CI	Action
0	0	ACR = min {ER, ACR + RIF × PCR, PCR}
0	1	ACR = min {ER, ACR(1 − RDF)}
1	0	ACR = min {ER, ACR}
1	1	ACR = min {ER, ACR(1 − RDF)}

When progressing the call, the network may, if necessary, adjust the ABR setup parameters as shown in Table 4.9.

The values of RDF may be increased or decreased as long as the ratio RDF/RIF is not decreased. That is, if RIF is decreased by a factor k, RDF may be decreased by at most a factor k, or it may be increased. Furthermore, the traffic parameter selection procedure is required to maintain the following invariant:

$$MCR \leq ICR \leq PCR$$

If the network is not able to provide PCR with a value greater than or equal to MCR, the call is cleared with cause "user cell rate not available." Fi-

Table 4.9
ABR Parameter Modification

Parameter for a Given Direction	Modification by the Network
PCR	Decrease only
ICR	Decrease only
TBE	Decrease only
RIF	Decrease only
RDF	Decrease/increase

nally, the network adjusts the fixed round-trip time parameter in the ABR setup parameters IE when forwarding a SETUP message for an ABR connection. The amount of the adjustment is the fixed portion of the RM cell delay through the network.

4.4.2.2 Call/Connection Establishment at the Destination Interface

Upon receiving a SETUP message, the called user examines the received ATM traffic descriptor and ABR setup parameters IEs. MCR is negotiated using the Q.2962 procedures if the corresponding minimum acceptable ATM traffic descriptor value for MCR is included in the SETUP message. To accept the call/connection, the user may take one of the following actions:

1. If the indicated PCR and ABR setup parameter values can be supported, the user accepts the call and returns a CONNECT message with the PCR, MCR, and ICR parameter values received in the SETUP message.

2. If the indicated PCR can not be supported but at least the MCR value can be provided, the user accepts the call and returns a CONNECT message after adjusting the PCR values as needed (the adjusted PCR value is greater than or equal to the MCR value).

   When progressing the call, the called user can, if indicated in the SETUP, also adjust the ICR parameter value and the following ABR setup parameters: ICR, TBE, RIF, and RDF. The called user may adjust the values of these parameters in either the forward or backward directions, or both.

3. If the user cannot support the MCR, the call establishment request is rejected and a RELEASE COMPLETE message is returned with cause "resources unavailable, unspecified."

4.4.3 Allowable Combination of Parameters in the SETUP Message

Table 4.10 shows the allowable combinations of the broadband bearer class, the ATM traffic descriptor parameters, the end-to-end transit delay, and the QoS classes for the ABR service.

Table 4.10
Allowable Combinations of Parameters for ABR

Bearer Capability			
Broadband Bearer Class	C	X	VP
Broadband Transfer Capability	Available bit rate		
Traffic Descriptor for a Given Direction			
PCR (CLP = 0)	Unspecified		
PCR (CLP = 0 + 1)	Specified		
SCR, MBS (CLP = 0)	Unspecified		
SCR, MBS (CLP = 0 + 1)	Unspecified		
ABR MCR	Optional in the user-network direction; mandatory in the network-user direction		
Tagging	No		
Frame discard	Yes/No		
ABR setup parameters	Specified		
QoS classes	0		

4.5 ATM Block Transfer Capability

An ABT capability is an ATM layer mechanism defined to provide a bearer service in which the ATM layer transfer characteristics are negotiated on an

ATM block basis. When a block of ATM cells is accepted, the network allocates sufficient resources to the block so that the QoS experienced by the ATM cells in a block is equivalent to the QoS provided by a DBR connection with the same PCR.

Two ABT traffic-handling capabilities are defined: ATM block transfer with delayed transmission (ABT-DT) and ATM block transfer with immediate transmission (ABT-IT).

In ABT-DT, the cell rate of successive ATM blocks is dynamically modified between the users of the ABT-DT capability and the network using resource management cells. Positive acknowledgment from the network is required before transmitting ATM cell blocks at the new cell rate.

In ABT-IT, the user can transmit a block of ATM cells without waiting for a positive acknowledgment from the network. As a result, a block of ATM cells may be discarded by the network if sufficient network resources are not actually available at that moment.

4.5.1　Signaling for ABT Transfer Capability

ABT transfer capability is supported using the Q.2931 messages and procedures with additional traffic parameters added to different messages for the support of the service.

An ATM traffic descriptor IE (see Figure 4.9) is included in the CONNECT message to indicate the traffic parameter values allocated for the call/connection if one or more traffic parameters in the SETUP message were negotiable. The ATM traffic descriptor IE is extended to include forward and backward RM PCR.

A minimum acceptable ATM traffic descriptor IE is included in the SETUP message when PCR (0 + 1) or RM PCR or both traffic parameters are negotiable. This IE is extended to include forward and backward RM peak cell rate.

Field	Octets
Forward RM peak cell rate identifier	21*
Forward RM peak cell rate	21.1 to 21.3*
Backward RM peak cell rate identifier	22*
Backward RM peak cell rate	22.1 to 22.3*

Figure 4.9　Extension of the ATM traffic descriptor IE for ABT.

4.5.2 Signaling Procedures

The Q.2931 procedures for basic call/connection control applies with indication of the SCR and MBS traffic parameters in ATM traffic descriptor IE. For the negotiation of the PCR, SCR, MBR, and RM peak cell rate, during the call establishment, the Q.2962 procedures are used. The additional procedures required to handle the point-to-point ABT call/connections are described next.

4.5.2.1 Call/Connection Establishment at the Originating Interface

If the calling party requests an ABT-DT (an ABT-IT) ATM transfer capability for the call/connection, the calling party sends a SETUP message containing a broadband bearer capability IE specifying ABT-DT (ABT-IT) in the broadband transfer capability field and point-to-point in the user plane connection configuration field. In addition:

- Tagging can not be requested.
- If the calling user wants to allow negotiation of a PCR (CLP = 0 + 1) parameter value, it includes the corresponding PCR parameter in the minimum acceptable ATM traffic descriptor IE.
- The calling party includes the forward and backward RM PCR parameters in the ATM traffic descriptor IE.
- If the calling user allows negotiation of the RM PCR value, it includes the corresponding RM PCR parameter in the minimum acceptable ATM traffic descriptor IE.

If the network or called user determines that the ATM traffic descriptor IE contains a nonsupported set of traffic parameters, a RELEASE COMPLETE message is returned with cause "unsupported combination of traffic parameters."

4.5.3 Allowable Combination of Parameters in the SETUP Message

The parameters specified in the broadband bearer capability IE, the ATM traffic descriptor IE, the end-to-end transit delay IE, and the QoS parameter IE of the SETUP message should be consistent. Table 4.11 shows the allowable combinations of these values.

The default values of SCR and MBS are equal to 0 and 1, respectively.

4.6 Frame Discard

If a network element needs to discard cells, it is in many cases more effective to discard the cells at the frame level rather than at the cell level. The term "frame"

Table 4.11
Allowable Combinations of Parameters for ABT-DT or ABT-IT

ATM Service Category	ABT-DT or ABT-IT			ABT-DT or ABT-IT		
	Bearer Capability					
Broadband Bearer Class	C	X	FR	C	X	FR
Broadband Transfer Capability	ABT-DT or ABT-IT			ABT-DT or ABT-IT		
	Traffic Descriptor for a Given Direction					
PCR (CLP = 0)	Unspecified			Unspecified		
PCR (CLP = 0 + 1)	Specified			Specified		
SCR, MBS (CLP = 0)	Unspecified			Unspecified		
SCR, MBS (CLP = 0 + 1)	Specified			Specified		
RM PCR	Specified			Specified		
Tagging	No			No		
Frame discard	No			No		
QoS classes	Network option			Network option		
Max. transit delay in the forward direction	Specified			Optional		

in this context is used to refer to an AAL protocol data unit. The network can detect the frame boundaries by examining the SDU-type in the payload-type field of the ATM cell header. If a network supports frame discard, it can treat user data as frames only if the user requests such treatment during call setup.

Upon receiving a SETUP message with the forward frame discard bit coded to "frame discard allowed," the network, if possible, passes this indication towards the called party and may perform frame discard for the forward direction of the connection.

Upon receiving an indication of backward "frame discard allowed," the network passes this indication towards the calling party in a CONNECT message and may perform frame discard in the backward direction of the connection.

The end system interprets the receipt of a SETUP message with the backward frame discard bit coded to "frame discard allowed" as a request by the calling party that the called end system requests backward frame discard in the CONNECT message. When the called user wishes to request frame discard for the backward direction, the user includes a backward frame discard indication coded to "frame discard allowed" in the CONNECT message.

4.7 Connection Characteristics Negotiation During Establishment Phase

This section describes the signaling protocol used to negotiate the connection characteristics of point-to-point calls/connections and for the first party of point-to-multipoint calls/connections. These negotiation capabilities are applicable only during the call/connection-establishment phase for DBR and SBR ATM service categories. In particular, the following capabilities are specified:

- Negotiation of connection characteristics using an alternative ATM traffic descriptor;
- Negotiation of peak cell rate traffic parameters using a minimum acceptable ATM traffic descriptor.

When the alternative ATM traffic descriptor IE is used, the parameters of the IE are handled as a single entity, whereas the minimum acceptable ATM traffic descriptor IE allows the specification of a range of values for parameters that are handled independently. Hence, the use of the alternative ATM traffic descriptor IE allows negotiation of all traffic parameters, whereas the use of the minimum acceptable ATM traffic descriptor is restricted to the negotiation of PCR and ABR MCR.

Either the alternative ATM traffic descriptor IE or the minimum acceptable ATM traffic descriptor IE (but not both) are included in the SETUP message when traffic parameters are negotiable. The ATM traffic descriptor IE is included in the CONNECT message if one or more traffic parameters were negotiable in the SETUP message (see Figure 4.10).

Field	Length
Alternative ATM traffic descriptor	4–20
Minimum acceptable ATM traffic descriptor	4–30

Figure 4.10 ATM traffic descriptor IE.

4.7.1   Information Elements

4.7.1.1   Alternative ATM Traffic Descriptor

The alternative ATM traffic descriptor IE is used to specify an alternative ATM traffic descriptor for the negotiation of traffic parameters during call/connection setup. This IE is applicable to DBR and SBR calls. This IE is coded the same as ATM traffic descriptor from octet 5 onwards.

The alternative ATM traffic descriptor IE can have any combination of traffic parameters allowed in the ATM traffic descriptor IE. Within a single SETUP message, the combination of traffic parameters may be different for these two information elements.

4.7.1.2   Minimum Acceptable ATM Traffic Descriptor

The minimum acceptable ATM traffic descriptor IE is used to specify the minimum acceptable ATM traffic parameters in the negotiation of traffic parameters during call/connection setup. The minimum acceptable ATM traffic parameters are the lowest values that the user is willing to accept for the call/connection. This IE is coded the same as the ATM traffic descriptor for octet groups 5–8.

4.7.2   Negotiating the Connection Characteristics

The user initiates the negotiation of the connection characteristics by including either the minimum acceptable ATM traffic descriptor IE or the alternative ATM traffic descriptor IE in the SETUP message, in addition to the ATM traffic descriptor IE. If a minimum acceptable ATM traffic descriptor IE is used, the indicated cell rates are lower than the corresponding cell rates specified in the ATM traffic descriptor IE.

If point-to-multipoint procedures are supported, the user may use the negotiation procedures only for the first party of a point-to-multipoint call. If the negotiation procedures are initiated, the user can send ADD PARTY messages only if the link is in the active state.

When both the minimum acceptable ATM traffic descriptor and the alternative ATM traffic descriptor IEs are present in a SETUP message, the call is rejected with cause "unsupported combination of traffic parameters."

4.7.2.1   Minimum Acceptable ATM Traffic Parameter Negotiation

If the network is able to provide the traffic parameter values specified in the ATM traffic descriptor IE, it progresses the connection-establishment request. If the network cannot provide some of the cell rates indicated in the ATM traffic descriptor IE but is able to provide at least their corresponding cell rates in the minimum acceptable ATM traffic descriptor IE, the connection-establishment request is progressed after the cell rates are adjusted in the ATM traffic descriptor IE.

If some of the parameters in the minimum acceptable ATM traffic descriptor IE are still less than the corresponding parameters in the modified ATM traffic descriptor IE, then the call/connection is progressed with the minimum acceptable ATM traffic descriptor IE containing all such parameters in addition to the modified ATM traffic descriptor IE (see Figure 4.11). Otherwise, the call/connection progresses with the modified ATM traffic descriptor IE and without the minimum acceptable ATM traffic descriptor IE.

4.7.2.2 Destination User

If the user is not able to provide some of the cell rates indicated in the ATM traffic descriptor IE, but is able to provide at least their corresponding cell rates in the minimum acceptable ATM traffic descriptor IE, the user progresses the connection-establishment request.

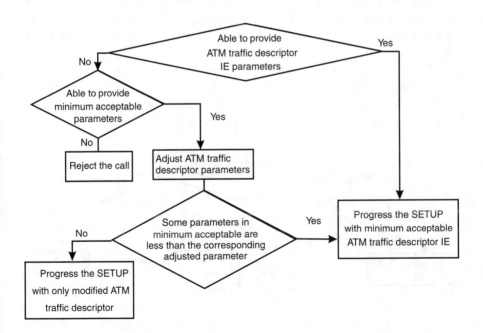

Figure 4.11 Network negotiation with minimum acceptable ATM traffic parameters.

4.7.2.3 Alternative Traffic Parameter Negotiation

If the network is able to provide the traffic parameter values specified in the ATM traffic descriptor IE, it progresses the connection establishment with the following options:

- Both the ATM traffic descriptor IE and the alternative ATM traffic descriptor IE, when the network can provide the traffic parameter values specified in the alternative ATM traffic descriptor IE;
- Only with the ATM traffic descriptor IE, if the network cannot support traffic parameter values specified in the alternative ATM traffic descriptor IE.

If the network cannot provide the ATM traffic descriptor indicated in the ATM traffic descriptor IE but is able to provide the ATM traffic descriptor indicated in the alternative ATM traffic descriptor IE, it progresses the connection-establishment request by using the contents of the alternative ATM traffic descriptor IE as the ATM traffic descriptor (see Figure 4.12).

If the user cannot provide the ATM traffic descriptor indicated by the ATM traffic descriptor IE, but is able to provide the ATM traffic descriptor indicated by the alternative ATM traffic descriptor IE, the user progresses the connection-establishment request on the basis of the alternative ATM traffic descriptor IE.

If the network or user can provide neither the ATM traffic descriptor indicated in the ATM traffic descriptor IE nor the ATM traffic descriptor indicated in the alternative ATM traffic descriptor IE, the connection-establishment request is rejected with cause "user cell rate unavailable."

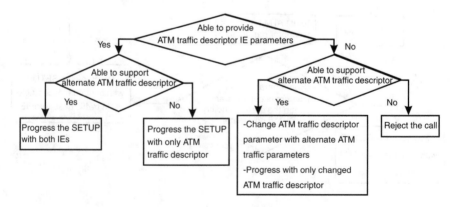

Figure 4.12 Network negotiation with the alternate ATM traffic descriptor IE.

4.7.3 Negotiation Acceptance

Upon receiving an indication that the request has been accepted, the network sends a CONNECT message across the UNI and enters the active state. The message returned to the user contains the ATM traffic descriptor IE indicating the cell rates finally allocated to the connection.

If no ATM traffic descriptor IE is included in the CONNECT message, the connection characteristics specified in the ATM traffic descriptor IE sent in the SETUP message applies.

If the user has progressed the call based on either the minimum acceptable ATM traffic descriptor IE or the alternative traffic descriptor IE, the message returned by the user contains the ATM traffic descriptor IE with the accepted connection characteristics.

5

B-ISDN Supplementary Services

The basic B-ISDN bearer service transports end-user data between two or more end stations. A supplementary service, on the other hand, enhances the basic service in different ways. In particular, each supplementary service is defined to add value to the underlying basic service. By definition, a basic bearer service must exist before a supplementary service can be provided.

Various supplementary services are already familiar to a majority of telephone users: call waiting and call forwarding are two examples. New supplementary services have come out with ISDN. B-ISDN supplementary services are being defined based on the ISDN supplementary services, but B-ISDN-specific new service has yet to be defined.

Supplementary services are defined in I.251 and I.257. Signaling support for some of these services are developed in Q.2951 and Q.2957. This chapter is an overview of the material provided in these ITU recommendations. Signaling procedures defined in Q.2951 for supplementary services include direct dialing in, multiple subscriber number, calling line identification presentation, calling line identification restriction, connected line identification presentation, connected line identification restriction, and subaddressing, whereas signaling procedures for user-to-user signaling are defined in Q.2957.

The current supplementary services are all address-based services. ITU-T is currently in the process of defining various other supplementary services (e.g., closed user group, call transfer, etc.). The addressed-based supplementary services are used to build computer-assisted applications. The information that is available from B-ISDN supplementary services can be passed to the computer systems supporting the application and database. The call-related information (e.g., calling line identification, subaddress, etc.) from the call is passed to the computer system. The application can incorporate call information into normal processing.

The applications that can use address based supplementary services are as follows:

- *Personal telephone support*—provides improved human interface and support for users;
- *Intelligent answering*—provides support for handling customer inquiries;
- *Telemarketing*—Provides support for marketing using telecommunications capabilities;
- *Integrated message desk*—integrates the telephone message desk facility with computer-based e-mail.

The link of computer and telephony is commonly referred to as intelligent answering. It uses the calling line identification presentation supplementary service. This information helps identify the caller and the nature of his or her request prior to the call being routed to a caller agent qualified to handle the request.

5.1 Direct Dialing In (DDI)

The DDI supplementary service enables a user to call directly to another user on an ISDN private branch exchange (ISPBX) or other private system without human intervention. In this context, a DDI number is a part of the ISDN number significant to an ISPBX or another private system. The DDI supplementary service is defined in I.251.1.

DDI is provided when at least a part of the ISDN number significant to the called user is passed to that user. These last digits (fixed or variable length) are received by the ISPBX or another private system, which automatically establishes a call to the destination without the assistance of an operator.

The Q.2931 call control procedures are used at the originating network side. At the destination network side, the DDI number is passed to the user in the called party number IE. The destination network side may check the range and the format of the DDI number before it is passed to the called user. The DDI number is delivered from the network to the called user using the Q.2931 procedures. If the addressing/numbering plan identification is "ISDN/telephony numbering plan (Recommendation E.164)," the type of the number is included in the called party number IE sent to the called user and it is coded as "unknown or subscriber number, national number or international number."

When this service is supported across an interface between a B-ISDN and an N-ISDN, the following simple mapping is defined. In the B-ISDN to ISDN direction, the B-ISDN called party number IE is mapped to the ISDN called party number IE by removing its second octet and adjusting the length indication without any other changes to the contents. Similarly, in the ISDN to B-ISDN direction, the ISDN called party number IE is mapped to the B-ISDN called party number IE by adding a second octet and adjusting the length indication field without any other changes to the contents.

5.2 Multiple Subscriber Number (MSN)

The MSN supplementary service allows the assignment of multiple ISDN numbers to an interface. This capability allows, for example, a calling user to select one or multiple distinct terminals or to identify the station that may provide other supplementary services described throughout this chapter.

The user's view of the MSN supplementary service description is defined in I.251.2 and the protocol requirements for this service at the UNI are defined in Q.2951.

The MSN may be a whole ISDN number or a part of the ISDN number. In the latter case, an MSN is the least significant n digit(s), where n is less than or equal to the full length of the ISDN number. The number of digits used for MSN should be large enough to assign a unique individual number to all stations attached to a network that are part of an administrative domain (organization). Alternatively, the MSN, as a service provider option, may be a number that can be mapped from an ISDN number by the network at the destination network side.

The Q.2931 call control procedures are used at the originating network side. If the MSN supplementary service is provided to the subscriber, the network may use (as a network option) the information in the calling party number IE to identify the calling party, and if necessary, assign the appropriate basic or supplementary service profile. At the option of the service provider, one of the MSN numbers may be designated by the MSN subscriber as the default number for the interface.

At the destination network side, the network sends the available part of the called party number or the corresponding number to the user in the Q.2931 SETUP message. The MSN of the called user is coded in the called party number IE, whereas the MSN of the calling user is coded in the calling party number IE.

The MSN, if provided by the calling user, is delivered from the user to the network using the Q.2931 procedures. If the addressing/numbering plan identification is "ISDN/telephony numbering plan (Recommendation E.164)" or "unknown," the type of the number indicated in the calling party number IE is sent to the network as "unknown" or "subscriber number," "national number" or "international number."

If the user sends an MSN, sufficient digits are supplied to uniquely identify an ISDN number from the set of ISDN numbers at that interface. If the network receives fewer digits in the calling party number IE than is required to uniquely identify an ISDN number at that interface, it discards this IE and behaves as if the calling party number IE was not received.

Similarly, the MSN is delivered from the network to the called user at the destination interface using the Q.2931 procedures. If the called user receives fewer digits in the called party number IE than it requires, it uses the available information in that IE for its terminal selection procedure.

The MSN service cannot be used when a private ISDN station is attached at the interface. Interworking in the N-ISDN to B-ISDN direction is not applicable. In the B-ISDN to N-ISDN direction, the called party number IE is mapped to the called party number IE by the interworking function by removing its second octet and adjusting the length indication without any other changes to the contents.

5.3 Calling Line Identification Presentation (CLIP)

The CLIP supplementary service provides the calling party's ISDN number to the called party, possibly together with the subaddress information. The provided information consists of the ISDN number of the calling user in a form sufficient to allow the called user to return the call (i.e., a subscriber number, a national number, or an international number, and, optionally, a subaddress if provided by the calling user). The CLIP service is defined in I.251.3.

The CLIP supplementary service is most commonly used service for computer-assisted applications. The calling party information is passed from the network to destination user and forwarded to the application. The call center makes it possible to synchronize the routing of the telephone call to a customer service representative telephone set with the presentation of related information to the agent's computer screen. The CLIP information helps to identify the caller and routed to a call agent qualified to handle the request. For example, specified accounts can be routed to selected agents for handling or to other departments (e.g., accounts receivable for those accounts that warrant special attention).

Companies that have implemented this capability have realized the productivity increase, as the time to handle a call is reduced. This is mainly due to a profile that appears at the workstation as the call arrives. Further, it enables the agent to commence business with the customer as soon as they answer the call, rather than having to first establish the caller's identity. The benefit of this capability is fewer agents handling more calls.

For callers in an automatic call distribution (ACD) queue that abandons the call without speaking to an agent, their CLIP information can be used to generate a callback. Such proactive calling can generate goodwill for callers who tire of waiting on hold. This information can be used to produce call logs that can be used for time tracking, billing, and other purposes.

The CLIP can be also used for personal telephone applications and small business applications. The voice mail can be stored and retrieved from personal computers using the calling party as index. The small business telephone order shop (e.g., auto parts) can automatically display information such as a customer sales profile, indexed in a database under the calling party number.

5.3.1 Actions at the Originating Switch

At the originating network side, all information pertaining to CLIP is included in the SETUP message sent as part of the Q.2931 call procedures (see Figure 5.1). If no information is provided by the calling user, the network provides the default number associated with the user interface at which the call was generated in the originating switch. In this context, the default number is an agreed upon number between the calling user and the network provider.

All information pertaining to CLIP is inserted in the SETUP message in the calling party number IE. The purpose of the calling party number IE is to identify the origin of a call, whereas the calling party subaddress IE is used to identify a subaddress associated with the origin of the call.

At the originating interface, the CLIP service may be provided as part of the basic service without the calling user being subscribed to the service. The addressing/numbering plan identification indicated in the calling party number IE sent by the calling user is either "ISDN/telephony numbering plan (Recommendation E.164)," "ATM endsystem address (NSAP address)," or "unknown." In the case of E.164 addresses, the complete subscriber number, national number, or international number is sent. If a partial calling number is included by the calling user (e.g., to indicate digits specific to the DDI or the MSN supplementary services), the user sets the type of number to "unknown."

If the calling user has not subscribed to the CLIP supplementary service, the network at the originating interface checks to see if the calling party number and calling party subaddress IEs are included in the SETUP message. If the calling party number IE is received with the addressing/numbering plan identification field coded anything other than "ISDN/telephony numbering plan

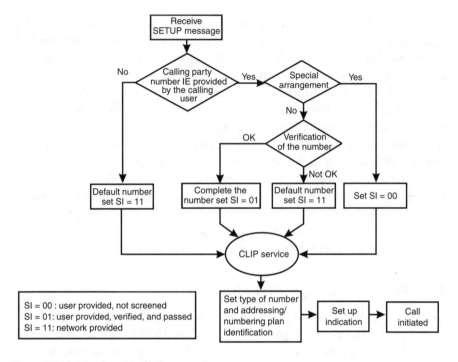

Figure 5.1 Originating side CLIP processing.

(Recommendation E.164)," "ATM endsystem address (NSAP address)," or "unknown," the network discards the calling party number IE and processes the call as if that information was not received. The network sets the value of the screening indicator based on the outcome of the screening operation of the calling number. The network disregards any screening indicator value, if received from the calling user. If the calling party number IE is included, the network performs the screening function.

If the calling number received from the calling user is determined to be correct, the network sets the screening indicator to "user-provided, verified and passed." If the screening function fails, the network indicates that the screening is failed and uses a default number associated with the calling user. In this case, the network sets the screening indicator to "network provided."

If the SETUP message does not contain the calling party number IE, the network uses a default number associated with the calling user. In this case, the network sets the screening indicator to "network provided."

If the calling user provides only partial calling number information and the number is a valid digit sequence for the user access arrangement, the network completes the number as appropriate and sets the screening indicator to "user-provided, verified and passed."

The calling user information is forwarded to the destination switch in association with the basic call request. The presentation indicator, as determined by the procedures of the calling line identification restriction (CLIR) supplementary service (described in Section 5.4), is forwarded to the destination switch, in association with the basic call request. If the calling party subaddress IE is available, it is passed transparently through the network.

The actions at the originating local exchange when special arrangement does not apply are summarized in Table 5.1.

Table 5.1
Information Provided by the Calling User and by the Network When Special Arrangement
Does Not Apply

Information Provided by the Calling User		Information Provided by the Network to the Called User		
Calling number received from the calling user	Type of number	Calling number forwarded if CLIR is not activated	Screening indicator forwarded	Type of number forwarded
No calling party number IE is provided by the calling user		Default number stored at the network side sufficient for returning the call	Network provided	International number or national number for E.164 Unknown for ATM endsystem address
Valid part of the number not sufficient for returning the call	Unknown	Completion of the number	User provided, verified and passed	International number or national number for E.164 Unknown for ATM endsystem address
Correct complete calling party number	Subscriber number, national number, or international number for E.164 Unknown for ATM endsystem address	Complete calling party number	User provided, verified and passed	International number or national number for E.164 Unknown for ATM endsystem address

Table 5.1. (Continued)

Information Provided by the Calling User		Information Provided by the Network to the Called User		
Calling number received from the calling user	Type of number	Calling number forwarded if CLIR is not activated	Screening indicator forwarded	Type of number forwarded
Incorrect number; the number provided by the user is discarded	Any type of number	Default number stored at the network side sufficient for returning the call	Network provided	International number or national number for E.164 Unknown for ATM endsystem address

If the destination is in a different country, a national number provided at the originating interface is converted to an international number by the public network. As a network option, the type of the number forwarded to the called user may be coded "unknown" when a prefix is added to the number. The prefixes or the absence of a prefix, is used to distinguish international numbers and national numbers from each other.

The actions at the originating switch when special arrangement applies are summarized in Table 5.2.

5.3.2 Actions at the Destination Switch

If the called user is provided with the CLIP supplementary service, the network checks to see if the calling number is available. If it is available and presentation is allowed according to the presentation indicator supplied in the calling number, the network includes the calling party number IE in the SETUP message sent to the called user. If provided, the network also includes the calling party subaddress IE in the SETUP message. The presentation and screening indicators associated with the calling number are passed transparently to the called user. The addressing/numbering plan identification field is coded either "ISDN/telephony numbering plan (Recommendation E.164)," "ATM endsystem Address (NSAP addressing)," or "unknown."

If the calling number or a "presentation restricted" indication is not available at the destination switch, the network includes the calling party number IE

Table 5.2
Information Provided by the Calling User and by the Network When Special Arrangement Applies

Information Provided by the Calling User		Information Provided by the Network to the Called User		
Calling number received from the calling user	Type of number	Calling party number forwarded if CLIR is not activated	Screening indicator forwarded	Type of number forwarded
No calling party number IE is provided by the calling user		Default number stored at the network side sufficient for returning the call	Network provided	International number or national number for E.164 Unknown for ATM endsystem address
Any digit sequence conforming to Rec. E.164	National number or international number	Number provided by the user	User provided, not screened	International number or national number for E.164 Unknown for ATM endsystem address

in the SETUP message sent to the called user. The presentation indicator is set to "number not available due to interworking," and the screening indicator is set to "network-provided," the type of number and the addressing/numbering plan identification is set to "unknown" and the number digits field is not included. The network does not include the calling party subaddress IE, if provided, in the SETUP message.

If the called user is not provided with the CLIP supplementary service, then neither the calling party number nor the calling party subaddress IEs are included in the SETUP message sent to the called user.

If presentation is restricted, but as a national network option the called user has the *override* category (e.g., police or emergency service) marked in the destination switch, the network includes the calling party number IE and calling party subaddress IE if the subaddress was supplied by the calling party in the

SETUP message. In this case, the presentation and screening indicators are passed transparently to the called user. Figure 5.2 illustrates CLIP processing at the destination switch.

5.3.3 Interworking

In the B-ISDN to N-ISDN direction, the B-ISDN calling party number IE and the calling party subaddress IE (if present) are mapped to the ISDN called party number IE by interworking function by removing their second octets and adjusting the length indication without any other changes to the contents. In the N-ISDN to B-ISDN direction, the N-ISDN calling party number IE and the calling party subaddress IE (if present) are mapped to the B-ISDN called party number IE by the interworking function by adding a second octet and adjusting the length indication without any other changes to the contents.

On calls incoming from a non-ISDN, the calling number may be delivered to the destination ISDN without an indication of calling line identity restriction. In this case, a number of options exist and the selection is based on the network rules and regulations:

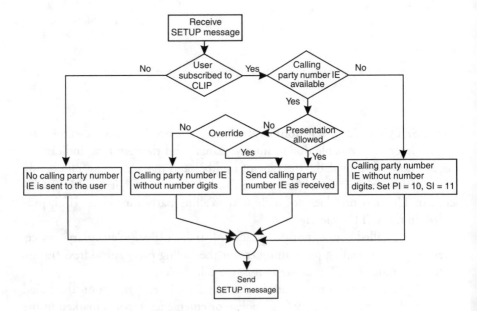

Figure 5.2 Destination switch CLIP processing.

- The network sends the calling party number IE; the calling party subaddress IE is not included.
- The network sends the calling party number IE and includes the calling party subaddress IE if the calling subaddress is available.

If the complete calling number is not available to the ISDN, the full number of the calling user cannot be given to the called user. In this case, the network sends the calling party number IE and does not include the calling party subaddress IE.

On calls incoming from some non-ISDN source, the calling number may be delivered to the destination ISDN with an indication of calling line identity restriction. In this case, the network sends the calling party number IE and includes the calling party subaddress IE if the calling subaddress is available.

As a network option, the originating network may restrict any address information identifying the calling user from being forwarded to another network. The calling line identification is not presented if the calling user has an arrangement to inhibit the presentation of his or her number(s) to the called party. The only occasion when a user subscribing to CLIP can take precedence over CLIR is when the CLIP-served user has an override category. This is a national option.

5.3.4 Two-Calling Party Number Information Elements Delivery Option

Q.2951 also specifies the procedures for an optional capability to deliver two-calling party number IEs to the called user. The CLIP procedures covers when the terminal is directly attached to the network. When the terminal is attached to a PBX or private network and the called party number is sent with screening indicator to the network, the network discards the screening indicator and performs the screening. The default number is provided when the verification is failed and special arrangements are not provided. This case is very common during public and private network interworking. By providing two numbers, the network can keep the user-provided number as a second number. This feature will enhance support of the intelligent answering capability. At present, this feature is not supported by all network equipment. These procedures are extensions of the CLIP procedures. In particular, the CLIP procedures are applied with no changes, when:

1. Only a single number is available for delivery at the terminating exchange;
2. Presentation is restricted;

3. The called user is not provided with the CLIP supplementary service;

4. If the subscription option to two-number delivery exists and the called user has not subscribed to two-number delivery.

When two numbers are available at the terminating exchange with screening indicators of one set to "network provided" and the other set to "user provided not screened," the network delivers the information in a *two-calling party number* IE sent in a SETUP message to the called user.

In addition, when two numbers are available at the terminating exchange with screening indicators of one set to "network provided" and the other set to "user-provided, verified and failed," the network delivers the information in a two-calling party number IE sent in a SETUP message to the called user.

If provided, the network also includes the calling party subaddress IE in the SETUP message. The presentation and screening indicators associated with the calling number, and the calling subaddress, received at the destination exchange, are passed transparently to the called user.

When the screening function fails, the network sets the screening indicator to "user-provided, verified and failed." The presentation indicator, as determined by the procedures of the CLIR supplementary service are forwarded to the destination local exchange in association with the basic call request.

If the calling party subaddress IE is available, it is passed transparently through the network.

The actions at the originating local exchange when screening fails are summarized in Table 5.3.

Table 5.3
Information Provided by the Calling User and by the Network When Screening Fails

Information Provided by the Calling User		Information Provided by the Network to the Called User		
Any digit sequence conforming to Rec. E.164	Unknown number, national number, or international number	Number provided by the user	User provided, verified and failed	Unknown number, national number, or international number
ATM endsystem address	Unknown number	Number provided by the user	User provided, verified and failed	Unknown number

5.4 Calling Line Identification Restriction (CLIR)

CLIR is a supplementary service offered to the calling party to restrict the presentation of the calling party's ISDN number and subaddress to the called party. When a CLIR is applicable and activated, the originating network provides the destination network with an indication that the calling user's ISDN number and subaddress (if provided by the calling user) are not allowed to be presented to the called user and the corresponding IEs are not included in the SETUP message sent to the called party. This feature, however, does not forbid forwarding of the calling party number within the network as part of the basic service procedures (cf. I.251.4).

At the originating network side, all information pertaining to the CLIR is inserted in the Q.2931 SETUP message. If the calling user wishes to override the default setting in the network, the SETUP message sent by the user may contain the calling party number IE with the presentation indicator set appropriately.

If the calling user has subscribed to the CLIR supplementary service on a permanent basis, the presentation indicator received in the SETUP message is ignored. The network sets the presentation indicator to "presentation restricted." If the calling user has subscribed to the CLIR supplementary service on a per-call basis and requests to override the default setting, the originating network sets the presentation indicator according to that in the received calling party number IE. If the CLIR is requested by the user on a per-call basis and no calling party number IE is included in the SETUP message, the originating network sets the presentation indicator according to the subscribed default value. The presentation indicator is forwarded to the destination with the basic call request.

The actions performed at the destination local exchange are provided as part of the CLIP service.

5.4.1 Interworking

In the B-ISDN to N-ISDN direction, the B-ISDN calling party number IE is mapped to the N-ISDN called party number IE by the interworking function by removing its second octet and adjusting the length indication without any other changes to the contents. In the N-ISDN to B-ISDN direction, the N-ISDN calling party number IE is mapped to the B-ISDN called party number IE by the interworking function by adding a second octet and adjusting the length indication without any other changes to the contents.

On calls to or via non-ISDN, it cannot be assured that a restriction indication is carried to the destination network. As a national network option, the originating network has the responsibility to restrict any information identifying the calling party being forwarded to the destination network when CLIR is applicable. If a destination network receives a calling party number without any indication of presentation allowed, or if it is restricted, the destination network (the host network) acts according to its own rules and regulations.

CLIR takes precedence over CLIP. The only occasion when a user subscribing to CLIP can take precedence over CLIR is when the CLIP-served user has an override category, which is a national option. The override category includes calls to police and other emergency services. Figure 5.3 illustrates CLIR procedures at the originating network side.

5.5 Connected Line Identification Presentation (COLP)

COLP is a supplementary service offered to the calling party. The connected party number is used to provide OSI X.213 services. It is also used to identify the connected party in case of call transfer and PBX-to-PBX connections. This service is used to provide the connected party's ISDN number with possible

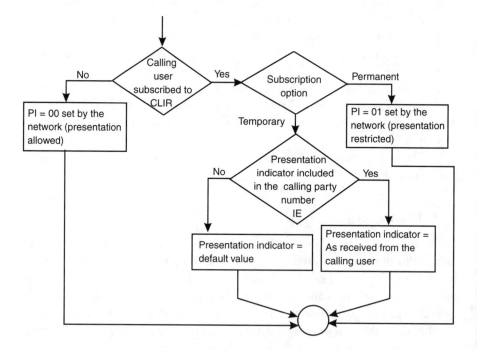

Figure 5.3 Originating network side CLIR procedure.

additional subaddress information to the calling party. This supplementary service is not used to check dialing. Instead, it is an indication to the calling subscriber of the connected ISDN number. In an ISDN environment, the connected line identity must include all the information necessary to unambiguously identify the connected line. The information may also include additional subaddress information (the connected subaddress IE) generated by the connected user. The network is not responsible for the content of this additional subaddress information (cf. I.251.5).

All information pertaining to COLP is inserted in the CONNECT message sent as part of the Q.2931 call procedures. If no information is provided by the connected user, the network provides the default number associated with this user's access at the destination switch. When the connected number is provided by the connected user, the network can only verify that the number is within the set of numbers allocated to that user. Where a special arrangement exists with the connected user, no verification is performed.

The COLP supplementary service makes use of the connected number and connected subaddress IEs inserted in the CONNECT message (see Figure 5.4).

5.5.1　Information Elements

5.5.1.1　Connected Number Information Element

The purpose of the *connected number* IE is to indicate the number connected to a call. The connected number may be different from the called party number because of changes (e.g., call redirection, transfer) during the lifetime of a call. The connected number IE is coded as shown in Figure 5.5. The maximum length of this information element is network dependent.

5.5.1.2　Connected Subaddress Information Element

The purpose of the *connected subaddress* IE is to identify the subaddress of the connected user of a call. The connected subaddress may be different from the called party subaddress because of changes (e.g., call redirection, transfer) during the lifetime of a call.

Information Element	Type	Length
Connected number	O	4–26
Connected subaddress	O	4–25

Figure 5.4 Information elements used for COLP service in the CONNECT message.

0/1 ext	Type of number			Addressing/Numbering plan identification	5
1 ext	Presentation indicator	0	0 Spare	0	Screening indicator	5a
0 ext	Number digits					6 etc.
ATM endsystem address						6 etc.

Figure 5.5 Connected number IE.

The connected subaddress IE is considered as an access information element, as described in Chapter 2. It is coded as shown in Figure 5.6. The maximum length of this IE is 25 octets.

5.5.2 Actions at the Originating Local Exchange

When the network sends a CONNECT message to the calling user and the calling user is provided with the COLP supplementary service, the network checks to see if the connected number is available. If it is available and presentation is allowed, the network includes the connected number IE in the CONNECT message sent to the calling user. If provided, the network also includes the connected subaddress IE. The presentation and screening indicators associated with the connected number received at the originating switch are passed transparently to the calling user.

The addressing/numbering plan identification field is coded either as "ISDN/telephony numbering plan (E.164)," "ATM endsystem address (NSAP addressing)," or "unknown." If presentation is not allowed according to the presentation indicator, the network includes the connected number IE in the CONNECT message sent to the calling user. The presentation indicator in the connected number IE indicates "presentation restricted." The network encodes the screening indicator, addressing/numbering plan identification, and the type of number according to one of the following options:

1 ext	Type of subaddress	Odd/even ind.	0	0 Spare	0	5
subaddress information						6 etc.

Figure 5.6 Connected subaddress IE.

1. The screening indicator indicates "network provided." The type of number and the addressing/numbering plan identification are set to "unknown."

2. The screening indicator, addressing/numbering plan identification, and the type of number are passed as received at the destination network.

If neither the connected number nor an indication that presentation is restricted is available at the originating switch, the network includes the connected number IE in the CONNECT message sent to the calling user. The presentation indicator is set to "number not available" and the screening indicator is set to "network provided," the type of number and addressing/numbering plan identification are set to "unknown," and the number digits field is not included. The network does not include the connected subaddress IE, if provided, in the CONNECT message.

If the calling user is not provided with the COLP supplementary service, then neither the connected number nor the connected subaddress IEs are included in the CONNECT message sent to the calling user. If presentation is restricted, but as a national network option the calling user has the override category (e.g., police or emergency service) marked in the originating switch, the network includes the connected number IE and connected subaddress IE if the subaddress was supplied by the connected user in the CONNECT message. In this case, the presentation and screening indicators are passed transparently to the calling user.

5.5.3 Actions at the Destination Switch

The called party need not have subscribed to the COLP supplementary service. The addressing/numbering plan identification indicated within the connected number IE is either "ISDN/telephony numbering plan (Recommendation E.164)," "ATM endsystem address (NSAP)," or "unknown."

If the connected number included by the connected user is complete and the addressing/numbering plan identification field is "ISDN/telephony numbering plan (Recommendation E.164)," the type of number to be indicated in the connected number IE is one of the following:

- *Subscriber number*—the complete subscriber number is sent;
- *National number*—the complete national number is sent;
- *International number*—the complete international number is sent.

If the connected number included by the connected user is complete and the addressing/numbering plan identification field is "ATM endsystem address (NSAP)," the type of number to be indicated in the connected number IE is "unknown."

If a partial connected number is included by the connected user (e.g., to indicate digits specific to the multiple subscriber number supplementary service), the user sets the type of number to be indicated within the connected number IE to "unknown."

5.5.3.1 Actions at the Destination Switch When a Special Arrangement Does Not Apply

Upon receiving a CONNECT message from the connected user, the network checks to see if the connected number and connected subaddress IEs are included. If the connected number IE is received with a coding of the addressing/ numbering plan identification field other than "ISDN/telephony numbering plan (Recommendation E.164)," "ATM endsystem address (NSAP)," or "unknown," the network discards the connected number IE and processes the call as if that information was not received.

The network sets the value of the screening indicator based on the outcome of the screening of the connected number. The network disregards any value of the screening indicator, if received from the connected user. If the connected number IE is included, the network performs the screening function. If the received connected number is determined to be correct, the network sets the screening indicator to "user-provided, verified and passed." If the screening function fails, the network notes that the screening is failed and uses a default number associated with the connected user and sets the screening indicator to "network-provided."

If the CONNECT message does not contain the connected number IE, the network uses a default number associated with the connected user and sets the screening indicator to "network-provided."

If the connected user provides partial connected number information and the number is a valid digit sequence for the user access arrangement, the network completes the number as appropriate and sets the screening indicator to "user-provided, verified and passed."

The information, as determined by the procedures above, are forwarded to the originating switch using the basic call response.

When a CONNECT message is received from the connected user, the network checks to see if the connected number and connected subaddress IEs are included.

If the connected number IE is received with a coding of the type of number field other than "national number," "international number," "ISDN/telephony number plan (Recommendation E.164)" or "unknown," the network discards the connected number IE and processes the call as if that IE was not received. The network disregards any value of the screening indicator, if received from the connected user.

The valid combination of type of number and addressing/numbering plan fields are shown in Table 5.4.

Table 5.4
Number and Addressing/Numbering Plan Fields

Type of Number	Addressing/Numbering Plan
National number or international number	ISDN/telephony numbering plan (E.164)
Unknown	ATM endsystem address

If the CONNECT message does not contain the *connected number* IE, the network uses a default number associated with the connected user and sets the screening indicator to *network-provided.*

The information, as determined by the procedures above, are forwarded to the origination local exchange using the basic call response.

The actions at the destination switch when special arrangement does not apply are summarized in Table 5.5, whereas the actions at the destination local exchange when special arrangement applies are summarized in Table 5.6.

5.5.4 Interworking

In the N-ISDN to B-ISDN direction, the ISDN connected number IE is mapped to the B-ISDN connected number IE by the interworking function by inserting the second octet and changing the length indication from 1 to 2 octets. The ISDN connected subaddress IE is mapped to the B-ISDN connected subaddress IE similarly. In the direction from B-ISDN to N-ISDN, the B-ISDN connected number IE is mapped to the ISDN connected number IE by the interworking function by removing its second octet and adjusting the length indication without any other changes to the contents. The B-ISDN connected subaddress IE (if present) is mapped to the ISDN connected subaddress IE similarly.

Table 5.5
Information Provided by the Connected User and by the Network When Special Arrangement
Does Not Apply

Information Provided by the Connected User		Information Provided by the Network to the Calling User		
Connected number received from the connected user	Type of number	Connected number forwarded if COLR is not activated	Screening indicator forwarded	Type of number forwarded
No connected number IE is provided by the connected user		Default number stored at the network side sufficient for returning the call	Network provided	National number or international number for E.164 Unknown for ATM endsystem address
Valid part of the number not sufficient for returning the call	Unknown	Completion of the number	User provided, verified and passed	National number or international number for E.164 Unknown for ATM endsystem address
Correct complete connected number	Subscriber number, national number, or international number for E.164 Unknown for ATM endsystem address	Complete connected number	User provided, verified and passed	National number or international number for E.164 Unknown for ATM endsystem address
Incorrect number	Any type of number	Default number stored at the network side sufficient for returning the call	Network provided	National number or international number for E.164 Unknown for ATM endsystem address

Table 5.6
Information Provided by the Connected User and by the Network When Special Arrangement Applies

Information Provided by the Connected User		Information Provided by the Network to the Called User		
Connected number received from the connected user	Type of number	Connected number forwarded if COLR is not activated	Screening indicator forwarded	Type of number forwarded
No connected number IE is provided by the connected user		Default number stored at the network side sufficient for returning the call	Network provided	National number or international number for E.164 Unknown for ATM endsystem address
Any digit sequence conforming to Rec. E.164	National number or international number	Number provided by the user	User provided, not screened	International number or national number
ATM endsystem address	Unknown number	Number provided by the user	User provided, not screened	Unknown

On calls destined for a non-ISDN, the connected number may be delivered to the originating ISDN without an indication of connected line identity restriction. In this case, a number of options exist, and the selection of an option depends on the network rules and regulations:

- The network sends the connected number IE and does not include connected subaddress IE.
- The network sends the connected number IE and includes the connected subaddress IE if the connected subaddress is available.

For a non-ISDN network, the complete connected number may not be available to the ISDN and the full number of the connected user cannot be given to the calling user who has been provided with the COLP supplementary service. In this case, the network sends the connected number IE and does not include the connected subaddress IE.

On calls destined to a non-ISDN network, the connected number may be delivered to the originating ISDN with an indication of connected line identity restriction. In this case, the network sends the connected number IE and includes the connected subaddress IE if the connected subaddress is available.

As a network option, the destination network may restrict any information identifying the connected user from being forwarded to another network. The connected line identification is not presented to the calling user if the connected user has subscribed to COLR. The only occasion when a user subscribing to COLP can take precedence over COLR is when the COLP user has an override category. This is a national option.

Figures 5.7 and 5.8 illustrate COLP processing at the destination and originating switches, respectively.

5.6 Connected Line Identification Restriction (COLR)

COLR is a supplementary service offered to the connected party to restrict presentation of the connected party's ISDN number and subaddress to the calling

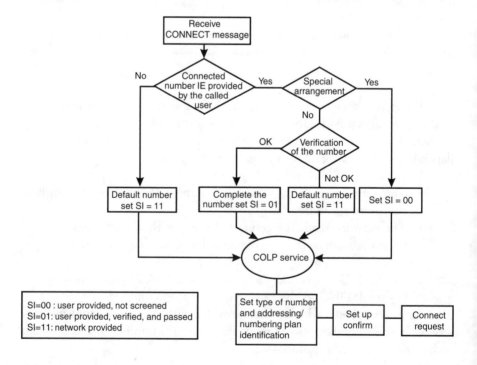

Figure 5.7 COLP processing at the destination switch.

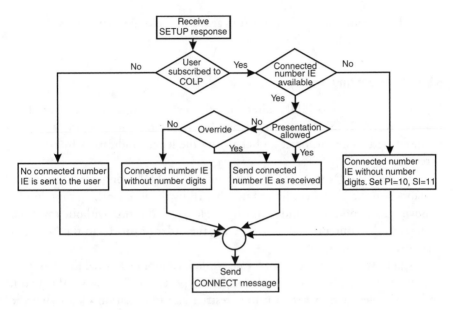

Figure 5.8 COLP processing at the originating switch.

party at the UNI (cf. I.251.6). When COLR is applicable and activated, the destination network provides the originating network with an indication that the connected user's ISDN number and subaddress information (if provided by the connected user) are not allowed to be presented to the calling user. In this case, no connected number and subaddress information is provided to the calling user in the CONNECT message.

All information pertaining to COLR is inserted in the CONNECT message sent as part of the Q.2931 call procedures.

The actions performed at the originating switch are provided as part of the COLP supplementary service. If the connected user wishes to override the default setting in the network, the CONNECT message sent by the user contains the connected number IE with the presentation indicator set appropriately.

If the connected user has subscribed to the COLR supplementary service on a permanent basis, the presentation indicator received in the CONNECT message is ignored. The network sets the presentation indicator to "presentation restricted." If the connected user has subscribed to the COLR supplementary service on a per-call basis and requests to override the default setting, the destination network sets the presentation indicator according to that in the received connected number IE.

If COLR is requested by the user on a per-call basis and no connected number IE is included in the CONNECT message, the destination network sets the presentation indicator according to the subscriber value.

The presentation indicator is forwarded to the originating network in association with the basic call response.

5.6.1 Interworking

In the N-ISDN to B-ISDN direction, the ISDN connected number IE is mapped to the B-ISDN connected number IE by the interworking function by inserting the second octet and changing the length indication from 1 to 2 octets.

In the B-ISDN to N-ISDN direction, the B-ISDN connected number IE is mapped to the ISDN connected number IE by the interworking function by removing its second octet and adjusting the length indication without any other changes to the contents and by respecting the order of this IE in the N-ISDN message.

On calls to or via a non-ISDN network, it cannot be assured that a restriction indication can be carried to the originating network. As a national network option, the destination network may restrict any information identifying the connected party from being returned to the originating network when COLR is applicable. If an originating network receives a connected party ISDN number without any indication of "presentation allowed" or "restricted," the originating network (the host network) acts according to its rules and regulations.

COLR takes precedence over COLP. The only occasion when a user subscribing to COLP can take precedence over COLR is when the user has an override category. This is a national option. COLR processing is summarized in Figure 5.9.

5.7 Subaddressing (SUB)

The SUB supplementary service allows the called user to expand his or her addressing capability beyond the one given by the B-ISDN number at the UNI, thereby providing an additional addressing capacity independent from the B-ISDN number. The applicability of the SUB supplementary service is specified in Recommendation I.251.8.

The functions offered by the SUB supplementary service can be used to identify a particular end point of a call beyond the B-ISDN access. If a calling user wants to transfer called party subaddress information to the called user, it inserts the called party subaddress information into the SETUP message. The subaddress information is transferred transparently through the network from the originating UNI to the destination UNI. At the called user side, the called

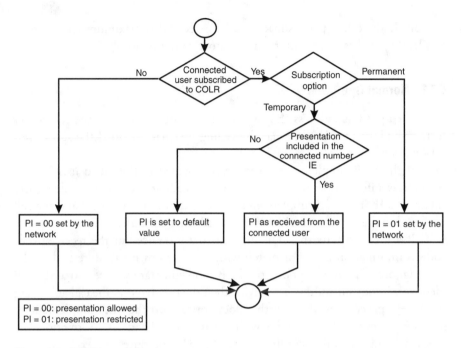

Figure 5.9 COLR processing.

party subaddress is offered to the called user within the SETUP message, if the called user has subscribed to this supplementary service.

Other subaddress IEs (e.g., calling party subaddress or connected party subaddress IEs) are not the subject of the SUB supplementary service. A called party subaddress, if presented by a calling user, is delivered unaffected to the called user. The called party subaddress may form part of the compatibility checking by the called user. Only the called user defines the significance of the subaddress.

The SUB supplementary service may be available without prior arrangement or it may be provided through a subscription agreement between the user and the service provider. If the subscription option is required, the user subscribes to the SUB supplementary service in order to receive called party subaddress information in incoming SETUP messages.

Withdrawal is done by the service provider at the subscriber's request or for administrative reasons. Both at the originating and the destination network sides, regular UNI signaling procedures apply.

For the SUB supplementary service, the calling user uses the called party subaddress IE. The maximum length of the called party subaddress information element is 25 octets, allowing for the transfer of 20 octets of subaddress information. If the called party subaddress IE exceeds this maximum length, then this IE is treated as an IE with nonmandatory content error.

5.7.1 Normal Operation

The called party subaddress IE is delivered from the network to the called user in the SETUP. This implies that the calling user has provided the subaddress information.

If the SUB supplementary service is provided to the called user but no subaddress information was included by the calling user in the called party subaddress IE, the SUB supplementary service cannot be provided and the call is offered to the called user without the called party subaddress IE.

If a user supports the SUB supplementary service but the received subaddress information does not match with the user's own subaddress, then the call is ignored. If a user supports the SUB supplementary service and a SETUP message without subaddress is received, then the user handles the call according to Q.2931 procedures. If a user that does not support the SUB supplementary service receives a SETUP message with called party subaddress information, the user will handle the call according to the Q.2931 procedures.

5.7.2 Interworking

In the B-ISDN to N-ISDN direction, the B-ISDN called party subaddress IE is mapped to the ISDN called party subaddress IE by the interworking function by removing its second octet and adjusting the length indication without any other changes to the contents. In the N-ISDN to B-ISDN direction, the N-ISDN called party subaddress IE is mapped to the B-ISDN called party subaddress IE by the interworking function by adding a second octet and adjusting the length indication without any other changes to the contents. If the call is not supported by the ISDN for the whole connection, the SUB supplementary service may not be applicable.

5.8 Relationship of Address Information Element and Supplementary Services

The correlation of address IEs to the basic call control or supplementary services is illustrated in Figure 5.10.

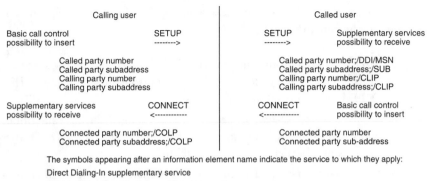

Figure 5.10 Correlation of address IEs to the basic call control or supplementary services.

5.9 User-to-User Signaling (UUS)

The UUS supplementary service, specified in Q.2957, allows a B-ISDN user to send/receive a limited amount of information to/from another B-ISDN user over the signaling virtual channel in association with a call/connection to the other B-ISDN user. In theory, it is possible for a calling user to use this service during the connection setup and pass a small amount of information to the called party. For example, the communicating entities may exchange some authentication protocol while the call is being established. The user to user data is up to 128 bytes and it is used for compatibility checking between calling and called party. It is also required for applications that are supporting OSI X.213 services and X.25. It is also used between PBX-to-PBX to pass additional information for the call. However, it is also a controversial service to the public providers. The called party receives this information and rejects the call setup request. How will the calling user be charged for this service? An alternative would be to include this charge in the base rate as a percentage of calls that are rejected.

The UUS supplementary services provide a means of communication between two users by using as a basis the layer 3 protocol. The exchange of user-to-user signaling is limited by flow control procedures provided by the network or the user. The exchange of user-to-user information is not a network-acknowledged service.

There are three UUS services associated with B-ISDN calls that may be provided by the network to users:

1. *Service 1*—User-to-user information is exchanged during the setup and clearing phases of a call/connection, by transporting the *user-user* IE in Q.2931 call/connection control messages.
2. *Service 2*—User-to-user information is exchanged during call/connection establishment, between the ALERTING and CONNECT messages, using USER INFORMATION messages.
3. *Service 3*—User-to-user information is exchanged while a call/connection is in the active state, using USER INFORMATION messages.

All three services may be used separately or in any combination during a single call. As an option, users may be able to specify that the requested user-to-user signaling service(s) is (are) required for the call during call/connection setup (i.e., the call should not be completed if user-to-user information cannot be passed).

Currently, only implicitly requested UUS service 1 is specified in Q.2957. Other services (i.e., explicitly requested UUS service 1, UUS service 2, and UUS service 3) are not defined. The calling user must subscribe to the implicitly requested UUS service 1 to use user-to-user signaling.

5.9.1 Messages

The following messages are applicable to the operation of the UUS service 1: SETUP, ALERTING, CONNECT, RELEASE, RELEASE COMPLETE, and PROGRESS. These signaling messages include the user-user IE when the calling user is subscribed to this service.

5.9.2 Information Elements

The purpose of the user-user IE is to convey information between B-ISDN users. This information is not interpreted by the network. It is carried transparently and delivered to the remote user. The length of the user-user IE is less than or equal to 133 bytes. The coding for the user-user IE is defined as shown in Figure 5.11.

Protocol discriminator (see Q.931 for coding)	5
User information	6–133

Figure 5.11 User-user IE.

5.9.3   UNI Signaling Procedures

The UUS service is provided using the Q.2931 procedures with the following modifications. Service 1 is implicitly requested by the calling user by including a user-user IE of variable length in the SETUP message and transferring it across the UNI. This IE is transported by the network and delivered unchanged in the user-user IE included in the SETUP message transferred across the UNI at the called side. For activation purposes, this IE must be at least 5 octets long (i.e., a minimum of 1 byte of user-to-user information needs to be included).

A user-user IE may be included in the ALERTING and/or CONNECT messages transferred across the UNI at the called side. It may also be included in the RELEASE or RELEASE COMPLETE message. The content of this IE is transported by the network and delivered in the user-user IE included in the corresponding message(s) transferred across the UNI at the calling side.

The network discards the user-user IE if it is received from the calling user in a SETUP message when the calling user has not subscribed to this service and continues to process the call request. The network also informs the calling user that the UUS request is not accepted by sending a STATUS message containing cause "requested facility nonsubscribed," or "access information discarded."

Similarly, the network discards the user-user IE if it is received from the called user in the ALERTING or CONNECT message, but a request for UUS was not indicated implicitly in the SETUP message delivered to the called user. If discard occurs, the network takes action on the remaining contents of the message received from the calling user and sends a STATUS message to the called user containing cause "access information discarded."

The called user may not be able to interpret incoming user-user IEs. In such situations, the user discards this information without disrupting normal call handling.

A user-user IE may be included in the first message used to initiate the normal call-clearing phase. The information contained in such an IE is transferred to the remote user in the first clearing message. This transfer is only performed if the information is received at the local switch of the remote user before sending a clearing message to that user; otherwise, the information is discarded without sending any notification. If the called user rejects the call setup request with a clearing message containing a user-user IE, the network delivers this IE in the RELEASE message sent to the calling user.

The network discards the user-user IE if it is received from either user in a RELEASE or RELEASE COMPLETE message and a request for UUS was not indicated implicitly in the SETUP message delivered to the called user. If discard occurs, the network takes action on the remaining contents of the message received from the user. If the clearing party has sent a RELEASE message, the network sends to the clearing party a RELEASE COMPLETE message containing cause "access information discarded." If the clearing party had sent a RELEASE COMPLETE message, the network considers the call as cleared to that party; no additional action is taken.

The network discards the user-user IE in the following cases:

- The overall length of the user-user IE is greater than 133 octets and UUS service 1 was activated implicitly;
- The network receives a message containing the user-user IE, but that message is not allowed to contain UUS.

If discard occurs, the network takes action on the remaining contents of the message received from the sending user and sends a STATUS message to that user containing cause "access information discarded." However, if the network discards a user-user IE from a received clearing message, it includes "access information discarded" in the next sequential clearing message sent to the user. If the network discards a user-user IE from a RELEASE COMPLETE message, it considers the call as cleared to that party and no additional action is taken.

5.9.4 Interworking

In the N-ISDN to B-ISDN direction, the ISDN user-user IE is mapped to the B-ISDN user-user IE by the interworking function by inserting octet 2 and changing the length indication from 1 to 2 octets. In the B-ISDN to N-ISDN direction, the B-ISDN user-user IE is mapped to the ISDN user-user IE by interworking function by removing its second octet and adjusting the length indication without any other changes to the contents.

In the case of interworking with a non-ISDN network, the return of a PROGRESS or an ALERTING message with the progress indicator IE indicating "call is not end-to-end ISDN; further call progress information may be available in-band" to the calling user serves as indication that the delivery of user-user IEs in call control messages cannot be guaranteed. Similarly, if the called user is not an ISDN user, the progress indicator IE indicating "destination address is non-ISDN" to the calling user serves as indication that the delivery of user-user IEs in call control messages cannot be guaranteed.

6

Interworking

For the first time in the history of networking, ATM-based B-ISDN provides the basic framework to be able to fulfill the promise of integrating all types of networking services and applications. Towards achieving this objective, B-ISDN should not only provide a framework to support emerging multimedia applications but also support services that are currently deployed. Two such services are the frame relay and narrowband ISDN (N-ISDN).

N-ISDN was developed to integrate speech, data, and various other services (such as telefax) based on 64-Kbps channels. Multiple channels may be used simultaneously to support higher bandwidths, up to 2 Mbps. When B-ISDN services become widely available, N-ISDN users will still want to be able to use some N-ISDN services. For example, a voice call that originates at a B-ISDN end station may terminate at a N-ISDN terminal.

The framework that defines the interworking between B-ISDN and N-ISDN has been specified by the ITU in I.580 and developed in Q.2931, section 6, entitled "Support of 64 Kbps Based Circuit-Mode services in B-ISDN Access Signaling Interworking Between N-ISDN and B-ISDN."

Frame relay is a fast packet technology that takes advantage of low bit error rates in the physical transmission medium and increasing intelligence and processing capacities at end stations. Frame relay eliminates much of the overhead and functions of the X.25 data-link layer. In particular, frame relay provides higher throughput and lower delays by eliminating resequencing, error correction, and retransmission within a frame relay network. Since its introduction, frame relay has proven to be a very successful wide area networking service. As ATM becomes increasingly deployed, the widely deployed frame relay networks and emerging ATM networks are expected to coexist for several years, requiring interworking between the two types of networks. The core aspects of frame relay are defined in Q.922, whereas signaling interworking between B-ISDN and frame relay is developed in Q.2933.

Interworking functions take into account the interworking of both signaling and user information. Interworking between ATM service and ISDN circuit-mode or frame relay service is performed by call control mapping. Signaling procedures provide the mechanisms for the negotiation of the user plane parameters. In order to allow signaling interworking between ATM and N-ISDN, and ATM and frame relay, various service-related information elements are defined in UNI signaling. To support ISDN circuit-mode services, bearer capability, high- and low-layer compatibility information elements of Q.931 are included in Q.2931. Similarly, the link layer core parameters and link layer protocol parameters information elements of Q.933 are included in Q.2931 to support the frame relay service.

6.1 ISDN Circuit Mode

The ATM signaling procedures to provision ISDN circuit mode are defined in Q.2931, section 6, providing the means to support 64-Kbps-based ISDN circuit capabilities. Figure 6.1 illustrates four scenarios that require circuit-mode ISDN and B-ISDN interworking.

In Figure 6.1(a), both end stations are connected to a B-ISDN and request an N-ISDN service (i.e., voice). The communication between the two end stations requires the mapping of both the user information and the signaling protocols of B-ISDN and N-ISDN.

One of the end stations in Figure 6.1(b) is connected to a B-ISDN, whereas the other is connected to a N-ISDN. The communication between these two end stations requires an interworking unit between the two networks.

In Figure 6.1(c), both end stations are connected to a B-ISDN and there is an N-ISDN backbone between the two networks. Similarly, in Figure 6.1(d), both end stations are connected to an N-ISDN and there is a B-ISDN between the two networks. In this case, B-ISDN is used as a backbone network. In both cases, interworking units are required for the communication to take place over two different types of networks.

In any one of these scenarios, the requested service may be a circuit-mode N-ISDN call. Hence, it is not possible to make any assumption about the network the called party is attached to. Accordingly, the calling user does not need to know or care how the connection is routed to the called user and whether the called user is connected to a B-ISDN or N-ISDN. This requires the use of the signaling interworking solution developed by ITU-T, which only requires

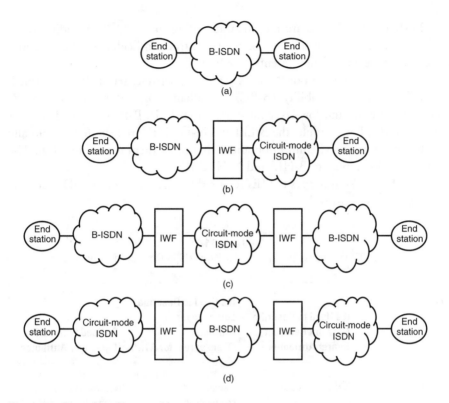

Figure 6.1 Examples of interworking.

B-ISDN users to indicate an N-ISDN service when they request a connection setup from a B-ISDN. Other than that, they do not need to know anything about the routing to and the location of the called party.

The interworking between N-ISDN and B-ISDN may take place between Q.2931 and Q.931 UNI signaling protocols or between ISUP and B-ISUP NNI signaling protocols. In this section, we will discuss only the UNI signaling internetworking procedures. ITU-T took a similar approach to address interworking at the NNI and the approach outlined next for UNI signaling interworking applies to NNI signaling interworking as well.

The basic approach taken by ITU-T is to emulate the N-ISDN services in B-ISDN, not to redefine them. This implies that the information elements defined for N-ISDN services in Q.931 are used in B-ISDN the same way, with no changes to their contents. In particular, to support ISDN circuit-mode services, bearer capability, high-layer compatibility and low-layer compatibility IEs of Q.931 are defined in Q.2931 as well. These IEs are designated as narrowband bearer capability (N-BC), narrowband low-layer compatibility (N-LLC) and narrowband high-layer compatibility (N-HLC). In order to reuse them in

Q.2931, the Q.931 IEs are modified according to the Q.2931 coding rules. That is, octet 2 of each IE contains the IE instruction field, and octets 3 and 4 indicate the length of the contents.

For the description of B-ISDN services, a separate set of IEs are defined: broadband bearer capability (B-BC), broadband high-layer information (B-HLI) and broadband low-layer information (B-LLI). B-ISDN services do not use these IEs. For example, the B-LLI and N-LLC IEs are regarded as mutually exclusive for an ATM connection and only one of them is used in a call. The same rule also applies to B-HLI and N-HLC IEs.

Table 6.1 summarizes the IEs required for the provision of ISDN circuit-mode services in B-ISDN.

Table 6.1
Q.2931 Information Elements Used to Provide ISDN Circuit-Mode Services

	IEs Used to Describe Network Relevant Bearer Attributes	IEs Used to Describe Lower Layer Attributes (Transparent for ATM)	IEs Used to Describe High-Layer Attributes
ISDN-related IEs	Narrowband bearer capability (N-BC)	Narrowband low-layer compatibility (N-LLC)	Narrowband high-layer compatibility (N-HLC)
ATM-related IEs	Broadband bearer capability (B-BC) ATM traffic descriptor QoS parameter End-to-end transit delay (optional) OAM traffic descriptor (optional)	ATM adaptation layer parameters	—

For N-ISDN services, the use of an N-BC IE is mandatory in the SETUP message. This IE is used to provide tones/announcements that are provided by the network, which happens only when the N-BC IE indicates "speech," "3.1 kHz audio," or "unrestricted digital information with tones and announcements." However, the N-BC IE is included in the SETUP message even if no interworking takes place, since the user cannot know in advance whether the receiver of the SETUP message is an ATM user or a circuit-mode ISDN user.

The inclusion of the ATM traffic descriptor IE is also mandatory in the SETUP message. N-ISDN services are based on 64-Kbps or $n \times$ 64-Kbps data rates. The ATM user cell rate needs to be selected such that the bit rate of the circuit-mode ISDN service can be transported as the cell payload of ATM cells. That is, the cell rate should be selected such that the ATM cell and the AAL header overhead are included in the signaled peak cell rate. Circuit-mode ISDN services use AAL 1. This is specified in the AAL parameters IE of the SETUP message as "circuit-mode ISDN services." AAL 1 cell payload is 47 bytes.

Table 6.2 illustrates the contents of the AAL parameters IE.

Table 6.2

AAL Parameters for Unrestricted Digital Information and Restricted Digital Information

Octet	Information Element Field	Field Value	
5	AAL-type	0000 0001	(AAL type 1)
6.1	Subtype	0000 0010	(circuit transport)
7.1	CBR rate	0000 0001	(64 Kbps)
9.1	Source clock frequency	0000 0000	(Null)
10.1	Error correction method	0000 0000	(Null)
11.1/11.2	Structured data transfer block size	0000 0000 0000 0000	(block size of 1)
12.1	Partially filled cells method	0000 0000	(Null)

Assuming all bytes of the cell payload are used, the peak cell rate of an ATM connection for a 64-Kbps circuit is $(53/47) \times$ 64 Kbps. More specifically, Table 6.3 illustrates the contents of the ATM traffic descriptor IE, assuming the use of AAL 1 with a 47-byte cell payload (i.e., partially filled cells are not currently allowed).

The quality of service parameter IE is always included in the SETUP message, and it is always set to "unspecified QoS class."

The N-LLC IE is used to describe the end-to-end attributes of ISDN circuit-mode services supported in ATM. If required for the description of a particular N-ISDN service, the N-LLC IE is included in the SETUP message and it is transported transparently through the ATM network. Similarly, if required for the description of a particular N-ISDN service, the N-HLC IE is included in the SETUP message. If included, it is transported transparently through the B-ISDN.

Table 6.3
ATM Traffic Descriptor for Unrestricted Digital Information and Restricted Digital Information

Octet	Information Element Field	Field Value if no OAM Cells Are Used	Field Value if 1 OAM Cell/s Is Used	Field Value With Maximal OAM Support
7.1	Forward peak cell	0000 0000	0000 0000	0000 0000
7.2	rate (CLP = 0 + 1)	0000 0000	0000 0000	0000 0000
7.3		1010 1011	1010 1100	1010 1111
		(171 cells/s)	(172 cells/s)	(175 cells/s)
8.1	Backward peak cell	0000 0000	0000 0000	0000 0000
8.2	rate (CLP = 0 + 1)	0000 0000	0000 0000	0000 0000
8.3		1010 1011	1010 1100	1010 1111
		(171 cells/s)	(172 cells/s)	(175 cells/s)

6.1.1 Interworking Between N-ISDN and B-ISDN

The signaling flows between N-ISDN and B-ISDN are almost identical. The only difference is the use of the RELEASE message in B-ISDN that corresponds to the DISCONNECT message in N-ISDN. The interworking function, in most cases, only needs to provide an appropriate mapping function for the IEs within a message.

In the N-ISDN to B-ISDN direction, the interworking function (IWF) maps the N-ISDN service-related information to B-ISDN service information and generates the additional IEs as needed. The N-ISDN bearer capability IE is mapped to the N-BC IE by simply inserting the second octet that contains the IE instruction field and changing the length indication from 1 to 2 octets. The flag bit in the second octet is set to 0. In addition to the N-BC IE, the B-BC IE is created by the IWF, indicating BCOB-A and the value "susceptible to clipping" in the susceptibility to clipping field. This prevents a possible clipping of voice information at the beginning of the communication. The ATM traffic descriptor and the QoS parameter IEs are generated by the IWF using the information in the N-ISDN BC IE. The N-ISDN LLC and HLC IEs (if included) are mapped to the N-LLC IE by the IWF without changing their contents. The AAL parameters IE is generated by the IWF, indicating AAL-type 1. Finally, the N-ISDN cause IE is mapped to the ATM cause IE by the IWF by simply inserting the second octet containing the IE instruction field and changing the length indication from 1 to 2 octets. The flag bit in the second octet is set to 0.

The mapping functions performed by the IWF for the direction from N-ISDN to B-ISDN are illustrated in Table 6.4. In this case, an N-ISDN user is requesting the unrestricted digital information bearer service.

Table 6.4
Mapping for the Unrestricted Digital Information Bearer Service (Direction N-ISDN to B-ISDN)

ISDN: Unrestricted Digital Information Bearer Service	B-ISDN: Emulation of the Circuit-mode ISDN Unrestricted Digital Information Bearer Service
BC: – Unrestricted digital information – Circuit mode – 64 Kbps	N-BC: – Unrestricted digital information – Circuit mode – 64 Kbps
HLC: Optional	N-HLC: Present, if provided
LLC: Optional	N-LLC: Present, if provided
—	B-BC: – BCOB-A – Susceptible to clipping – Point-to-point
—	ATM traffic descriptor
—	Quality of service: – Forward - Unspecified QoS class – Backward - Unspecified QoS class
—	AAL parameters

In the B-ISDN to N-ISDN direction, the B-BC, ATM traffic descriptor, QoS parameter, end-to-end transit delay, AAL parameters, and OAM traffic descriptor IEs are discarded by the IWF. The N-BC IE is mapped to the N-ISDN BC IE by simply removing its second octet and adjusting the length indication without any changes to its contents. Similarly, the N-LLC and N-HLC IEs (if included) are mapped to the N-ISDN LLC and HLC IEs by simply removing their second octets and adjusting the length indication without causing other changes to the contents.

The mapping functions performed by the IWF for the direction from ATM to ISDN are illustrated in Table 6.5. In this case, a B-ISDN user is requesting the N-ISDN unrestricted digital information bearer service.

Table 6.5
Mapping for the Unrestricted Digital Information Bearer Service (Direction B-ISDN to N-ISDN)

ATM: Emulation of the N-ISDN Unrestricted Digital Information Bearer Service	ISDN: Unrestricted Digital Information Bearer Service
N-BC: – Unrestricted digital information – Circuit mode – 64 Kbps	BC: – Unrestricted digital information – Circuit mode – 64 Kbps
N-HLC: Optional	HLC: Present, if provided
N-LLC: Optional	LLC: Present, if provided
B-BC: – BCOB-A – Susceptible to clipping	—
ATM traffic descriptor: Equal to 64 Kbps	—
Quality of Service: Unspecified QoS class	—
AAL parameters: AAL type 1	—
End-to-end transit delay: (optional)	—
OAM traffic descriptor: (optional)	—

The B-ISDN cause IE is mapped to the N-ISDN cause IE as shown in Table 6.6.

Table 6.6
ATM and ISDN Cause Values

ATM		N-ISDN	
35 36 37 45	Requested VPCI/VCI not available VPCI/VCI assignment failure User cell rate not available and No VPCI/VCI available	47	Resource unavailable, unspecified
73 93	Unsupported combination of traffic parameters AAL parameters cannot be supported	79	Service or option not implemented, unspecified

6.1.2 Notification of Interworking

When an emulated N-ISDN service is requested in B-ISDN, it is possible that the called party is not a B-ISDN or a N-ISDN terminal [i.e., plain old telephone service (POTS) phone]. In this case, N-ISDN indicates this information in a *progress indicator* IE. Hence, interworking between ATM and N-ISDN requires the support of this IE by the Q.2931. Unlike non-ISDN interworking with N-ISDN, a call leaving or entering the B-ISDN to/from the N-ISDN at an IWF does not in itself cause the generation of the progress indicator IE.

All progress indicator values that apply to N-ISDN interworking with non-ISDN have to be relayed through the B-ISDN to allow for indication to the calling user. The progress indicator IE is transported transparently through the B-ISDN. The ISDN progress indicator IE is mapped to/from Q.2931 by adding or removing octet 2, depending on the direction of the call, and adjusting the coding of the length indication.

6.1.3 Tones and Announcements

Users of N-ISDN services are used to certain service characteristics, with tones and announcements being among them. In order to provide the same services provided in N-ISDN, B-ISDN is required to provide tones and announcements during the call establishment. As discussed previously, this service is provided only if N-BC IE is present in the SETUP message indicating either "speech," "3.1 kHz audio," or "unrestricted digital information with tones and announcements."

The user need not attach to the virtual channel until receiving a CALL PROCEEDING, PROGRESS, or ALERTING message with the progress indicator "in-band information" or "appropriate pattern now available" or "call is not end-to-end ISDN: further call progress information may be available in-band." Prior to this time, the network cannot assume that the user has attached to the virtual channel. Upon receipt of the CONNECT message, the user attaches to the virtual channel (if it has not already done so).

The tones and announcements inserted by the network (B-ISDN or N-ISDN) are conveyed back to the calling user just like normal voice information, using the connection established for the voice service.

When in-band tones and announcements are provided, the RELEASE message sent by the network contains the progress indicator "in-band information or appropriate pattern now available." The network initiates clearing by sending the RELEASE message, starting timer T306, and entering the release indication state. Since in-band tones and announcements are being sent, even though the user connection is being released, the user may still connect to the virtual channel upon receiving a RELEASE message with this progress indicator. The user enters the release indication state upon receiving the RELEASE message. Alternatively, when the user does not want to receive in-band tones and announcements, it may continue clearing the connection by releasing the virtual channel and the call reference, sending a RELEASE COMPLETE message, and returning to the null state.

It is also possible that the user may subsequently continue clearing before receiving a RELEASE COMPLETE message from the network, even if connected to the in-band tone/announcement. In this case, the user releases the user virtual channel and the call reference, sends a RELEASE COMPLETE message, and returns to the null state.

Upon receiving the RELEASE COMPLETE message, the network stops timer T306, releases both the user virtual channel and the call reference, and returns to the null state. If timer T306 expires, the network continues clearing by releasing the user virtual channel and the call reference, sending a RELEASE COMPLETE message with the cause number originally contained in the RELEASE message, and returning to the null state. In addition to the original clearing cause, the RELEASE COMPLETE message may contain a second cause IE with cause "recovery on timer expiry." This cause may contain a diagnostic field identifying that the timer that expired.

6.2 Frame Relay Service

Frame relay (FR) is a connection-oriented data service, requiring a connection to be established between two FR stations prior to data flow. Frame relay supports access speeds of 56 Kbps, $n \times 64$ Kbps, 1.544/2.048 Mbps (T1/E1), and 44.7 Mbps (T3). The FR core services are described in Annex C of I.233.1 and are standardized in Q.922. The FR core aspects include the definition of frame formats and their fields and congestion management methods. Figure 6.2 illustrates the FR header formats.

The data-link connection identifier (DLCI) specifies separate connections across each access link between the end station and the access node on the frame relay network. The D/C bit is referred to as the DLCI or DL-CORE control indication. DL-CORE bits are used to transfer I.233 specific information (such as service attributes, performance criteria, etc.). The use of the C/R (command/response) bit is application specific. A forward explicit congestion notification (FECN) bit is used by the network to indicate congestion to the destination end point of the network. Similarly, a backward explicit congestion notification (BECN) bit is used by the network to notify the originator of the network congestion. A BECN bit may also be used by the destination end sta-

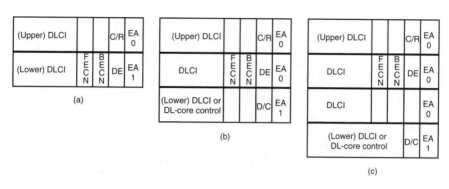

Figure 6.2 Frame relay (a) 2-, (b) 3-, and (c) 4-byte header formats.

tion to warn the originating end of network congestion. Finally, a discard eligibility (DE) bit can be set to indicate low-priority frames that can be discarded in the network in the event of network congestion.

The frame relay core services are provided in ATM on top of AAL 5 common part convergence sublayer by the frame relay service-specific convergence sublayer (FR-SSCS) specified in I.365.1. The frame relay UNI signaling procedures used to establish and release switched frame relay connections are defined in Q.933. Q.2933 specifies extensions to Q.2931 to emulate frame relay services over ATM networks. The states, information elements, messages, and procedures in Q.2933 are the same as in Q.2931 with minor extensions related to the frame relay service. In particular, the two Q.933 IEs, namely the link layer core parameters and link layer protocol parameters, are defined in Q.2933 to emulate frame relay over ATM. These two IEs are modified according to the ATM coding rules. That is, octet 2 is added as the IE instruction field and octets 3 and 4 are added to indicate the length of the contents. The emulated frame relay service has the following characteristics:

- It provides a bidirectional transfer of service data units from calling user to called user. Frames are delivered to the called user in the order they are transmitted by the calling user.
- There is a one to one correspondence between a DLCI and a VPI/VCI.
- The user plane procedures use the services provided by AAL type 5 common part on a VCC basis with special frame relay service-specific coordination function.

Recommendation I.555 defines two network interworking scenarios. In scenario 1, illustrated in Figure 6.3, two frame relay networks or frame relay end stations are connected via a B-ISDN. FR-SSCS specified in ITU-T Recommendation I.365.1 above the AAL type 5 common part is used to provide frame

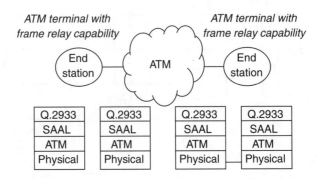

Figure 6.3 An ATM terminal with frame relay capability.

relay service over ATM. The approach taken by ITU is to emulate FR bearer service in an ATM end station (connected to an ATM network) and to provide interworking between an ATM network and an FR network. The ATM connection used by the frame relay service is a class C variable bit rate service.

In scenario 2, an ATM end station with a frame relay capability communicates with a frame relay end station connected to a frame relay network, as illustrated in Figure 6.4.

The IWF at the user plane (i.e., for the data traffic) maps the following features between the FR service functions and ATM cell relay functions:

- Variable length PDU formatting and delimiting;
- Error detection;
- Loss priority mapping (between the discard eligibility bit of the FR and the CLP bit of ATM cell);

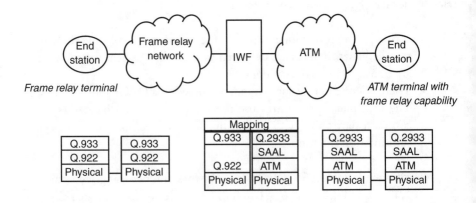

Fig 6.4 Frame relay to ATM interworking.

- Congestion indication mapping (between the EFCI bit of the FR and the EFCI bit of ATM cell).

The protocol stack on the signaling plane is shown in Figure 6.4. The interworking function will terminate both Q.933 and Q.2933 protocols. It also provides mapping function between the two protocols. Table 6.7 provides the mapping between frame relay and B-ISDN information elements when frame relay service is emulated over B-ISDN.

Table 6.7
Information Elements Used to Emulate Frame Relay in ATM

	Information Elements Used to Describe Network Relevant Bearer Attributes	Information Elements Used to Describe Lower Layer Attributes (Transparent for ATM)	Information Elements Used to Describe High-layer Attributes
Frame relay-related IE	Link layer core parameters	Link layer protocol parameters	—
ATM-related IE	Broadband bearer capability (B-BC) ATM traffic descriptor QoS parameter End-to-end transit delay	ATM adaptation layer parameters Broadband low-layer information	Broadband high-layer information

6.2.1 Mapping of Call Setup Messages

Q.933 and Q.2933 are both based on Q.931 signaling specifications and use the same call control messages and call control states. One major difference between Q.933 and Q.2933 is in the specification of traffic and QoS parameters. In the case of frame relay, the traffic parameters are included in the *link layer core parameter* IE, which specifies the maximum frame size, throughput, committed burst size, and excess burst size. In ATM networks, the traffic parameters are indicated in the ATM traffic descriptor and include the maximum burst size, peak cell rate, and sustainable cell rate.

When a Q.933 IE is mapped to the corresponding Q.2933 IE, it is modified to include the compatibility indication field (i.e., octet 2, flag, reserved bit, IE action identifier fields—see Chapter 1 for details). This capability allows enhancements in the feature and indicates what type of action will be taken when an IE is not recognized or has content errors. Currently, this field is set to 0 (i.e., to its default setting). Since each B-ISDN IEs contain the length of its contents (bytes 3 and 4), these FR-related IEs are extended to include a 2-byte length indicator field.

For frame relay service, the Q.933 IEs link layer core parameters and link layer protocol parameters are optionally included in the Q.2931 SETUP and CONNECT messages, even if no interworking takes place. When included, they are carried transparently by the ATM network.

The broadband bearer capability IE was modified to include a new code point in a bearer class (octet 5) and optionally new octet 7 to include a layer 2 identifier. This provides easy mapping of bearer capability information elements.

6.2.2   Traffic Parameters Mapping

The frame relay traffic parameters include committed information rate (CIR), committed burst size (Bc) and excess burst size (Be). The parameters may be different in each direction of the connection. The CIR describes the information transfer rate to which the network must commit in order to support a user during a normal data transfer. It is also referred to as throughput in Q.933. Associated with CIR is a time interval (Tc) during which a user can send up to the committed amount of data (Bc) and the excess amount of data (Be). The throughput, then, is the average number of "frame relay information" bits transferred per second in a given direction. The relationship between these parameters is as follows:

- If CIR > 0 and Bc > 0 and Be = 0, then Tc = Bc/CIR;
- If CIR = 0 and Bc = 0 and Be > 0, then Tc = Be/access rate of the link;
- If CIR > 0 and Bc > 0 and Be > 0, then Tc = Bc/CIR and EIR = Be/Tc where EIR is the excess information rate.

The ATM traffic parameters are defined in terms of PCR, SCR, and MBS. PCR specifies the maximum rate at which cells can be submitted to the network. The SCR is an upper bound on the possible conforming "average rate" of an ATM connection.

Traffic mapping between frame relay and ATM is provided by the IWF. In the frame relay to ATM direction, the PCR of the ATM connection can be derived from the throughput (CIR) and excess burst. The values of PCR and SCR are chosen to include an extra margin required to accommodate the overhead introduced in transferring the FR frames via an ATM network. The MBS for CLP = 0 traffic is set to allow the maximum burst Bc of the FR service to be provided at the PCR in ATM. Similarly, the MBS of the CLP = 1 traffic is set to allow the excess burst of the FR service to be provided at the PCR in ATM.

The ATM overhead occurs as a result of two factors:

- Overhead due to the 5-octet ATM cell header that is added to each 48 octets of the frame relay user data;
- Unused space in the last cell when a frame relay user data plus 8 octets (AAL 5 overhead) is not a multiple of 48.

Due to the nature of the ATM overhead, it is not possible to map the frame relay traffic parameters to ATM traffic parameters exactly. In particular, the ATM overhead depends on the packet length distribution. As two different packet length distributions with the same mean may have different ATM overhead, it is necessary to know the packet length distribution of frame relay traffic (which is almost never available) for the accurate computation of the ATM traffic parameters.

6.2.3 Interworking

A frame relay/ATM interworking unit can be viewed as having two sides: the frame relay protocol stack and the ATM protocol stack. In the frame relay to ATM direction, it terminates the frame relay protocol, provides a mapping function, and starts the ATM protocol (it is similar in the ATM to frame relay direction). The mapping function for signaling involves the following:

1. Frame relay signaling IEs that are used in ATM signaling (e.g., calling party number, calling party subaddress, transit network selection) are modified according to the ATM coding rules: octet 2 specifies the IE instruction field, whereas the length of the contents are indicated in octets 3 and 4. In addition, the required mapping is provided for the new IEs (e.g., low-layer compatibility, high-layer compatibility, and repeat indicator).
2. Information contents are mapped from one protocol to another based on the specifics of the mapped protocol.
3. IEs that are significant to only one network (e.g., protocol discriminator, call reference, and connection identifier, etc.) are deleted and no mapping takes place.

In ATM signaling, the inclusion of the link layer core parameters and link layer protocol parameters are optional. When they are included in an ATM

SETUP message, the IWF need not provide any mapping of traffic parameters. If they are not included, the mapping is similar to the one performed in the direction from frame relay to ATM, as discussed in Section 6.2.2. Table 6.8 shows the mapping of IE contents.

Table 6.8
Association of Various Parameters Between Q.933 and Q.2933

Q.933	Q.2933
Bearer capability	Broadband bearer capability
Link layer core parameters - Outgoing maximum frame size - Incoming maximum frame size	AAL parameters (AAL 5) - Forward maximum CPCS-SDU size - Backward maximum CPCS-SDU size
Link layer core parameters - Throughput - Committed burst size - Excess burst size	ATM traffic descriptor - Peak cell rate CLP = 0 + 1 - Sustainable cell rate CLP = 0 - Maximum burst size CLP = 0
Link layer core parameters - Minimum acceptable throughput	Minimum acceptable ATM traffic descriptor
End-to-end transit delay - Cumulative transit delay - Requested transit delay - Maximum transit delay	End-to-end transit delay - Cumulative transit delay - Requested transit delay - Not supported

7

Private NNI Signaling

Private network-network interface (PNNI) is a demarcation point between two switching systems. In this context, a switching system may be a single ATM switch or it may be a collection of two or more ATM switches managed and operated under one administration. In the former case, PNNI is a switch-to-switch interface, whereas it is a network-to-network interface in the latter case. In this chapter, we use the terms switch and switching system interchangeably. When used as an NNI, the protocol used internal to a network may not be a PNNI, in which case the PNNI is used as the interface to the outside world. As an example, a public network may use B-ISUP internally while communicating with the private networks attached to it via PNNIs. Depending on the information exposed by the public network across these PNNIs, this would allow private networks to make better decisions (i.e., path selection) when they communicate among themselves across a public network.

After a brief review of the PNNI framework, this chapter describes the signaling protocol used at the PNNI. Various features included in PNNI signaling include the specifications of the following:

- Signaling AAL used at the PNNI, including the specifications of service-specific coordination function, service-specific connection-oriented protocol, and AAL-5 common part convergence sublayer, and segmentation and reassembly sublayer;
- Point-to-point and point-to-multipoint call control procedures for switched virtual connections, switched virtual path connections, and soft permanent virtual path connections/soft permanent virtual channel connections;

- Crankback procedures used to reroute calls that are rejected within a PNNI domain;
- Designated transit list (DTL) processing to determine the downstream switch(es) along the end-to-end path;
- Quality of service framework.

7.1 An Overview of the PNNI Framework

ATM end stations are attached to a PNNI network across UNIs. Let us consider end station 1 (ES-1) attached to network A requesting a connection to end station 2 (ES-2) attached to network C, as illustrated in figure 7.1.

ES-1 sends across its UNI a SETUP message to switch A.1 providing the network with the information related to its connection request. In order for two or more end stations to communicate with each other, it is necessary to find a path in the network that has enough resources to support and manage the connection. The PNNI framework provides a standard-based solution to routing and signaling. PNNI signaling is used to establish, manage, and terminate connections across a PNNI, whereas the PNNI routing framework provides a set of basic capabilities for switches to find a path across the network between two end stations (point-to-point connection) or two or more end stations (point-to-multipoint connection).

In PNNI routing, the switching system that a connection request originates across its UNI (originating switch) is responsible for finding the end-to-end path to the destination end station(s). This is referred to as source routing. To determine the path, the originating switch uses link state routing in which each switch in the network advertises information about the PNNI links at-

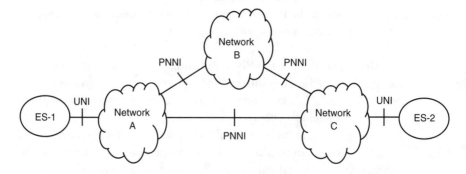

Figure 7.1 A PNNI network.

tached to it to other switches. In this context, a PNNI link connects one switch to another (in a given direction) across a PNNI. Both PNNI links and switches in this context could be physical network elements or logical elements. A logical switch may correspond to two or more switches grouped together. Similarly, a logical link may be used to connect two (logical) switches while crossing through one or more physical links in the network.

For each connection request, the originating switch finds a path based on the advertised capabilities and the desirability of other switching systems to carry connections with different characteristics. After finding the path that is likely to support the connection, the originating switching system uses PNNI signaling to request a connection establishment from intermediate switches along the selected path. The sequence of switching systems visited is specified in a DTL stack. Each switch processes connection request messages it receives, makes connection admission decisions (i.e., accept or reject the connection request), and passes the signaling message to the next switch along the path (if accepted) or denies the connection request and sends a clearance message to the preceding switch.

The choice of what internal state information to advertise, how often, and to where is based on a multilevel hierarchical routing model. The PNNI hierarchical model explains how each level of hierarchy works, how multiple nodes at one level are summarized into the higher layer, and how state information among nodes within the same level and between different levels are exchanged. The model is recursive: the mechanisms used to summarize the lowest level of PNNI routing into the next higher layer are the same as the mechanisms used to summarize any layer to the next higher layer.

At each level of hierarchy, the topology is represented by logical nodes and logical links. At the lowest level of the hierarchy, each node may represent a physical switch. At higher layers, each node may represent either a physical system or a group of switches. Consequently, PNNI links providing connectivity between the nodes may be either a physical link or a logical link.

Nodes are collected into peer groups. All the nodes within a peer group exchange link information among them and obtain an identical topology database representing the peer group. Peer groups are organized into a hierarchy in which one or more peer groups are associated with a parent peer group. Parent peer groups are grouped into higher layer peer groups, and so forth. The steps of forming a PNNI hierarchy are given as follows:

1. At the lowest level, switches are arranged into peer groups. One node in each peer group becomes the peer group leader.

2. Each peer group leader is responsible for specifying the parent peer group, either through configuration of the peer group leader or by a default peer group identifier based on a portion of its address.

3. Step 2 is repeated until all nodes are under one leader.

As an example, let us consider the network in Figure 7.2, taken from ATM Forum PNNI specification version 1.0 document. At the lowest level, the administration of this network decides to organize it, say, into seven peer groups A.1, A.2, A.3, B.1, B.2, B.3, and C. Each peer group has an identifier that ranges in length from 0 to 13 bytes (i.e., 0 to 104 bits), elects a peer group leader, and is assigned a parent level (indicated by the length of the identifier). Each peer group is also responsible for specifying the parent peer group. This may be accomplished by configuring nodes that are eligible to become the peer group leader or by using a default parent group identifier based on a prefix of the child peer group identifier.

Continuing with the example, the peer group leaders of A.1, A.2, and A.3 at level 96 discover that they have a common parent peer group identifier and form a single peer group at level 72 (the level number of the parent peer group is less than any of the peer group identifiers) called A. Similarly, peer groups B.1, B.2, B.3 form a single peer group B at level 80. Peer groups A, B, and C discover that they all belong to the same peer group at level 64. The resulting peer groups are illustrated in Figure 7.3.

Each node has the full description of its peer group's topology, including all its nodes, links, destinations that can be reached, and the status of various resources in the peer group. Routing outside a peer group follows the same link state operation, but it is achieved at the higher layers of the hierarchy. Considering a peer group as a logical node, topology information is exchanged between the logical nodes identifying each such node at the same hierarchy (i.e., level) and the logical links that connect them. An advertised link may correspond to a

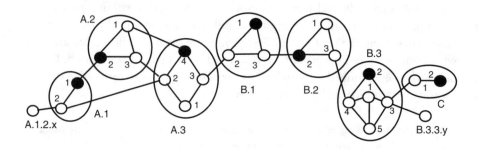

Figure 7.2 A PNNI network.

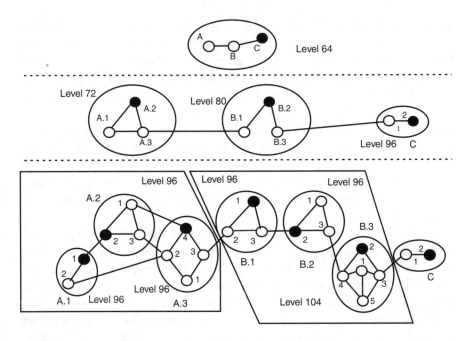

Figure 7.3 Hierarchical peer groups.

multihop path between real systems or may correspond to a link between logical group nodes. Each logical group node may represent an entire peer group.

The operation of PNNI routing in a parent group attempts to collapse a child peer group into a single node. In this case, there would be different paths to cross a peer group, each with different characteristics. Hence, it may not be always possible to advertise the true cost of the real physical paths. The representation of peer groups and their interconnections as logical nodes and logical links implies that information distributed at higher layers is summarized. This allows PNNI routing to scale to very large networks. Its consequence is that somewhat imperfect paths may be selected.

The list of end systems that are reachable through a logical group node is generally the complete list of systems reachable in the lower level child peer group that it represents. This information is summarized by address prefixes. Summarization of end system reachability is recursive. That is, those addresses announced at one level can further be summarized into more inclusive summary addresses at higher levels (with shorter prefixes). If there are additional end systems that cannot be summarized this way, they are advertised explicitly.

The hierarchical summarization discussed so far allows those switches that take part in the highest level of routing to calculate routes to any destination represented in the highest level peer group. However, it is necessary for all switches in the overall PNNI network to be able to route calls to any destination. This requires the higher level routing information to be available at the lower level switches. All switches participating in PNNI routing maintain link state databases for their peer groups, parent peer group, grandparent peer group, and so forth, up to and including the top level. This allows routes to be calculated on demand by any switch.

In order for the PNNI hierarchical routing to work properly, it is required to be used everywhere. In practice, however, there will be routes that do not support this protocol (e.g., public networks). To address this problem and utilize these links in PNNI routing, the concept of external routes is introduced. Each node advertises its special external routes, if any, to a particular set of destinations.

PNNI routing framework also includes the specification of a protocol for a switch to exchange messages with its immediate neighbor PNNI switches to determine its local topology. All nodes within a peer group exchange link state update (LSU) messages with each other to report their local topologies to all the others. These LSUs are exchanged reliably. Several other types of LSUs are used to allow a switch to announce its links to neighbor nodes, the metrics associated with each link, and the end systems reachable via the node. Some of these may not need to be exchanged reliably.

"Link state parameter" is a generic term that includes both link metrics and link attributes. A link metric requires its values for all the links along a given path to be combined to determine whether the path is acceptable and/or desirable for carrying a particular type of connection (such as maximum cell transfer delay and maximum cell loss ratio). A link attribute is considered individually for each link to determine whether a given link is acceptable and/or desirable to carry a particular type of connection (such as performance-related or policy-related attributes).

Using its topology, an originating switch finds a path to the destination that is likely to support the connection. The originating switching system then requests this path to be established from all switches (physical and/or logical) along the path using PNNI signaling. To ensure that the selected path is used, PNNI signaling includes a DTL in the call setup request. Each DTL contains the path elements for one sequence. Hence, the end-to-end path is specified as a DTL stack consisting of one or more DTLs.

Consider the network illustrated in Figure 7.2. Let us assume that a connection setup is requested from end station A.1.2.x to the end station B.3.3.y. A.1.2.x sends a UNI SETUP message to its switch A.1.2. A.1.2 examines its view of the world. The destination is reachable from peer group B. Examining the topology, A.1.2 finds that there are two paths to B: (A.1.2, A.1.1, A.2, A.3, and B) and (A.1.2, A.3, and B). Let's assume that the first path is chosen. Then, A.1.2 builds three DTLs in a stack:

DTL-1: {A.1.2, A.1.1}, destination pointer-2

DTL-2: {A.1, A.2, A.3}, destination pointer-1

DTL-3: {A, B}, destination pointer-1

Each DTL lists nodes that the call setup request needs to visit at a given hierarchical level. The destination pointer following each DTL specifies which element in the list is the next node to be visited at that level. When the end of the top DTL is reached, it is removed from the call setup request and the next DTL is examined.

Based on this framework, A.1.2 forwards the call setup request to its neighbor A.1.1, which looks at the top DTL, notices that the destination pointer points to itself, tries to advance it, finds it is exhausted, and removes the top DTL. The current destination is A.2 (after the pointer is moved to the next entry in the DTL). Since A.1.1 is not in the peer group summarized into the logical node A.2, it starts looking to see how to get to A.2. It finds that node A.2.2 is its immediate neighbor in A.2, so the call setup is sent to that node after removing the top DTL and advancing the destination pointer to 2. The new DTL stack now looks:

DTL-1: {A.1, A.2, A.3}, destination pointer-2

DTL-2: {A, B}, destination pointer-1

A.2.2 looks at the top DTL and sees that the current destination is A.2. Since A.2.2 is in A.2, it looks at the next entry in DTL and starts routing to A.3. Analyzing the topology, it finds out that the path is through A.2.3, so it pushes a new DTL onto the stack and sends the set up message to A.2.3:

DTL-1: {A.2.2, A.2.3}, destination pointer-2

DTL-2: {A.1, A.2, A.3}, destination pointer-2

DTL-3: {A, B}, destination pointer-1

A.2.3 determines that the top DTL target has been reached and the current DTL is exhausted. It then notices that the next node in the DTL is A.3, one of its neighbors. A.2.3 removes the top DTL and advances the current transit point to the next DTL:

DTL-1: {A.1, A.2, A.3}, destination pointer-3

DTL-2: {A, B}, destination pointer-1

A.3.2, upon receiving the setup message, notices that it is in A.3, the current node pointed to in the top DTL, and the target is reached. That leaves the target B. A.3.2 builds a route to B through A.3.4 and pushes a new DTL on the stack:

DTL-1: {A.3.2, A.3.4, A.3.3}, destination pointer-2

DTL-2: {A.1, A.2, A.3}, destination pointer-3

DTL-3: {A, B}, destination pointer - 1

A.3.2 then forwards the setup request to A.3.4, which repeats the process and forwards the request to A.3.3. Similarly, A.3.3 processes the request, notices that the DTL is exhausted, and the destination is now B. It determines that its neighbor B.1.2 is in B:

DTL-1: {A, B}, destination pointer-2

B.1.2 receives the setup and sees that the current DTL destination has been reached. B.1.2 must now build a new source route to the final destination. The path will go through B.2 to B.3. It then calculates the path through B.1. The setup message is sent to B.1.3 with the DTL stack:

DTL-1: {B.1.2, B.1.3}, destination pointer-2

DTL-2: {B.1, B.2, B.3}, destination pointer-1

DTL-3: {A, B}, destination pointer-2

B.1.3 forwards the setup request to B.2.2. B.2.2 processes the request, finds a path to B.3 through B.2.1 and B.2.3, and sends the request to B.2.1:

DTL-1: {B.2.2, B.2.1, B.2.3}, destination pointer-2

DTL-2: {B.1, B.2, B.3}, destination pointer-2

DTL-3: {A, B}, destination pointer-2

B.2.2 forwards the setup request to B.2.1, which looks at the top DTL, advances the pointer, and forwards the request to B.2.3. B.2.3 removes the DTL, forwards the setup request to its neighbor in B.3. B.3.4 builds a new DTL:

DTL-1: {B.3.4, B.3.1, B.3.3}, destination pointer-2

DTL-2: {B.1, B.2, B.3}, destination pointer-3

DTL-3: {A, B}, destination pointer-2

Proceeding in a similar fashion, the setup request eventually reaches the last switch along the path (i.e., B.3.3), which has a reachability to the destination end station. The last switch then starts running the UNI signaling procedures to establish the connection to the destination end station.

PNNI signaling allows calls to be established in either direction. In view of this, a reference model is used for a PNNI link based on the direction of the call: the preceding side and the succeeding side. The preceding side routes an outgoing call over the PNNI link and the succeeding side receives the incoming call. Similarly, the forward direction is the direction from the calling user to the called user, whereas the backward direction is the direction from the called user to the calling user. This framework is illustrated in Figure 7.4.

Figure 7.5 summarizes the relationship between the signaling procedures and underlying services within the control plane. PNNI signaling protocol specifies the procedures to dynamically establish, maintain, and clear ATM con-

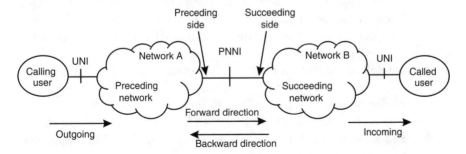

Figure 7.4 The PNNI reference model.

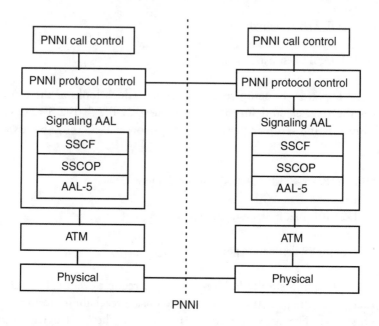

Figure 7.5 PNNI control plane.

nections at a PNNI between two ATM networks or two ATM switches. The PNNI call control entity performs PNNI routing and call admission-related functions. The PNNI protocol control (PNNI signaling) entity provides services to the PNNI call control by processing the actual signaling finite state machines for both incoming and outgoing calls.

PNNI signaling procedures use the services of the UNI signaling ATM adaption layer (SAAL). The SAAL layer comprises the following:

- A service-specific coordination function (SSCF) as specified in ITU-T Recommendation Q.2130. This function maps the service of the service-specific connection-oriented protocol (SSCOP) to the needs of the signaling procedures.
- SSCOP as specified in ITU-T Recommendation Q.2110. It is a peer-to-peer protocol providing assured services to transfer signaling information between any pair of SSCOP entities.
- The ATM adaptation layer type 5 (AAL-5) CPCS and segmentation and reassembly (SAR) sublayers as specified in ITU-T Recommendation I.363. This service provides the segmentation and reassembly of signaling data.

An assured mode connection between two peer signaling AALs is established before signaling messages can start flowing across a PNNI using SSCF and SSCOP. The details of these two layers are discussed in Chapter 1 in the context of UNI signaling.

The ATM layer, specified in ITU-T Recommendation I.150, provides for the transparent transfer of fixed size ATM layer service data units between communicating (signaling) AAL entities. The structure of the cell header used in the PNNI is the cell header format and encoding at the NNI. Any bits of the VPI or VCI subfield that are not allocated are set to 0. The physical layer can be any one of the physical layers specified for ATM by ATM Forum and other standards organizations. The specifications of the ATM layer and various physical interfaces defined for ATM can be found in related national and international standards, ATM Forum interoperability agreements, and various books written on ATM.

7.2 PNNI Signaling

Both PNNI point-to-point and point-to-multipoint signaling frameworks are extensions of ATM Forum UNI 3.1, UNI 4.0, and related ITU-T recommendations on UNI signaling. The procedures for point-to-point signaling are based on Q.2931, whereas the procedures for point-to-multipoint are based on Q.2971. The UNI specification distinguishes between the user side and the network side of the interface. In point-to-point PNNI signaling, call states and procedures are symmetric and they are based on the network-side counterparts at the UNI. In the case of point-to-multipoint PNNI signaling, in general, the network side of the UNI corresponds to the preceding side and the user side of the UNI corresponds to the succeeding side.

7.2.1 An Overview of PNNI Signaling Messages

The PNNI signaling message format is identical to that of Q.2931 messages (described in Chapter 1) and consists of the following fields: protocol discriminator; call reference; message type (including message compatibility instruction indicator); message length; and variable length information elements, as required.

The PNNI signaling protocol uses various messages to dynamically manage ATM connections across the PNNI as listed in Tables 7.1 and 7.2. A PNNI message is said to have a local significance if it is relevant only at the local PNNI, or it has a global significance meaning that it is relevant at the local

PNNI, other PNNIs along the connections path, and/or UNIs associated with the call.

Table 7.1
PNNI Signaling Messages for Point-to-Point Call Control

Message Type	Significance	Direction	Description
ALERTING	Global	Succeeding to preceding	Indicates that the called user alerting has been initiated
CALL PROCEEDING	Local	Succeeding to preceding	Indicates that the requested call/connection establishment has been initiated
CONNECT	Global	Succeeding to preceding	Indicates that the call/connection request is accepted by the called user
RELEASE	Global	Both	Indicates that the connection has cleared and the sender is awaiting to release the call reference
RELEASE COMPLETE	Local	Both	Indicates that the connection is internally cleared and the call reference is released
SETUP	Global	Preceding to succeeding	Initiates a call/connection establishment
STATUS	Local	Both	Sent in response to a STATUS ENQUIRY message or at any time to report certain error conditions
STATUS ENQUIRY	Local	Both	Solicits a STATUS message
NOTIFY	Access	Both	Indicates information pertaining to a call/connection

A point-to-point call/connection provides connectivity between two end stations attached to the network. In general, a point-to-point connection is bidirectional. However, by setting the traffic parameters (i.e., peak cell rate) to 0 in one direction, it could also be a unidirectional connection. Accordingly, PNNI point-to-point signaling messages are used across one or more PNNIs to establish a call/connection between two UNIs.

Table 7.2
PNNI Signaling Messages for Point-to-Multipoint Call Control

Message Type	Significance	Direction	Description
ADD PARTY	Global	Preceding to succeeding	Used to add a new party to an existing connection
ADD PARTY ACKNOWLEDGE	Global	Succeeding to preceding	Indicates that the ADD PARTY request was successful
PARTY ALERTING	Global	Succeeding to preceding	Indicates that the called user alerting has been initiated in response to an ALERTING message
ADD PARTY REJECT	Global	Succeeding to preceding	Used to indicate that the ADD PARTY request was not successful
DROP PARTY	Global	Both	Used to drop a party from an existing point-to-multipoint connection
DROP PARTY ACKNOWLEDGE	Local	Both	Indicates that the party is dropped from the connection

A point-to-multipoint call/connection provides a unidirectional connectivity from a single end station (root) to two or more end stations (leaves). Traffic flows only from the root to leaves. Point-to-multipoint PNNI signaling messages are used to add new leaves to an already established point-to-multipoint call/connection and terminate a connection to a leaf across a PNNI.

7.2.2 An Overview of Point-to-Point Call States

Similar to the other signaling protocols, the PNNI signaling protocol is described using a call state machine. Figure 7.6 illustrates the state machine used for managing point-to-point calls. The states defined for point-to-point call control are listed in Table 7.3.

The originating switch is required to know ahead of time that the connection is point-to-multipoint. This information is specified in the *broadband bearer capability* IE, the user plane connection configuration field. After the point-to-point connection with a point-to-multipoint indication is set up and the point-to-point call state is active, the addition/release of other leaves causes multipoint substates under the active call state to change. These states are listed in Table 7.4.

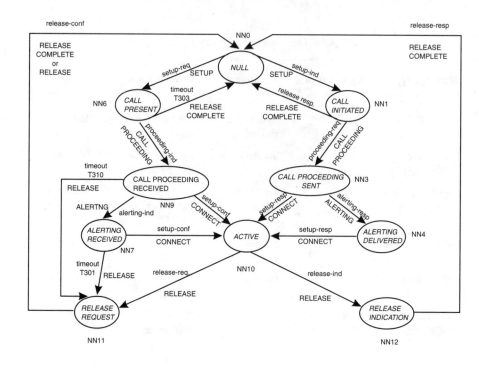

Figure 7.6 PNNI point-to-point state diagram.

Table 7.3
Point-to-Point Call States

State	Description
Null (NN0)	No call exists
Call Initiated (NN1)	This state exists in a succeeding side after it has received a call establishment request from the preceding network node but has not yet responded
Call Proceeding Sent (NN3)	This state exists when a succeeding side has acknowledged the receipt of the information necessary to establish a call
Alerting Delivered (NN4)	This state exists when a succeeding side has received an indication that the called user is alerting and sent an ALERTING message

State	Description
Call Present (NN6)	This state exists for a network node after it has sent a call establishment request to the succeeding side but has not yet received a response
Alerting Received (NN7)	This state exists when a preceding side has received an ALERTING message from the other side of PNNI interface
Call Proceeding Received (NN9)	This state exists when a preceding side has received acknowledgment that the succeeding side has received the call establishment request
Active (NN10)	This state exists when the ATM connection has been established
Release Request (NN11)	This state exists when a network node has sent a request to the other network node to release the ATM connection and is waiting for the response
Release Indication (NN12)	This state exists when a network node has received a request from the other network node to release the ATM connection and has not responded yet

Table 7.4
Point-to-Multipoint Call States

Substate	Description
Null	The party does not exist, no endpoint reference value has been allocated
Add party initiated	A SETUP or ADD PARTY message has been sent to the other side of the interface for this party of the call
Add party received	A SETUP or ADD PARTY message has been received from the preceding side for this party of the call
Drop party initiated	A DROP PARTY message has been sent for this party of the call
Drop party received	A DROP PARTY message has been received for this party of the call
Active	CONNECT, CONNECT ACKNOWLEDGE, or ADD PARTY ACKNOWLEDGE message identifying the party has been received

7.3 An Overview of PNNI Connection Setup

Before PNNI signaling messages can be exchanged across a PNNI, a signaling channel is required to exist between the preceding and succeeding sides. Two cases are allowed in the PNNI specification: associated signaling and nonassociated signaling (see Figure 7.7).

In associated signaling, the signaling entity exclusively selects and controls the VCs in the VPC to carry its signaling messages. Associated signaling procedures are used only when two PNNI network nodes are connected by a VPC used as a logical link.

In nonassociated signaling, the signaling entity controls all connections across a PNNI except the provisioned VPCs. The virtual channel with (virtual path identifier) VPI = 0/VCI = 5 is the only one used in PNNI for nonassociated signaling. PNNI is required to support the nonassociated signaling procedures and may as an option support the associated signaling procedures.

7.3.1 A Point-to-Point Call Setup

Initially, the call state is null. Upon receipt of a setup indication from the UNI (at the originating PNNI switch), the preceding side sends a PNNI SETUP message, enters the call present state, and starts awaiting for the CALL PROCEEDING message. The SETUP message has all the information required to set up the connection (i.e., traffic parameters, service requirements, called party address, etc.).

When the preceding side of a switch receives a setup indicator from its succeeding side, as illustrated in Figure 7.8, the procedures are the same as if the setup indication is received from a UNI. The only exemption is the process of

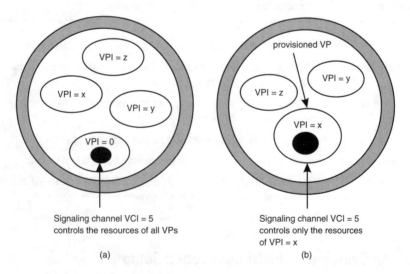

Figure 7.7 Signaling channels across a PNNI: (a) associated signaling and (b) nonassociated signaling.

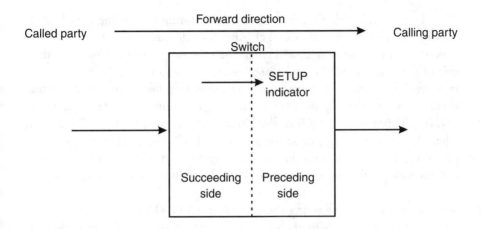

Figure 7.8 Preceding/succeeding sides of a switch.

DTL generation and DTL processing, as described in Section 7.5.4. In particular, the originating PNNI node is required to find a path in the network to the destination end station (i.e., DTL generation), whereas an intermediate node determines only the next node along the path from the designated transit list prepared by the originating switch (i.e., DTL processing).

Upon receiving a SETUP message, the succeeding side acknowledges it by sending a CALL PROCEEDING message to the preceding side. It then enters the *call proceeding sent* state. When the preceding side receives the CALL PROCEEDING message, it enters the *call proceeding received* state.

The succeeding side may receive an indication that the called party is alerting. Upon receiving this indication, it sends an ALERTING message to the preceding side and enters the *alerting delivered* state. Upon receiving the ALERTING message from the succeeding side, the preceding side enters the *alerting received* state.

Upon receiving an indication from call control that the call has been accepted, the succeeding side sends a CONNECT message to the preceding side and enters the active state. This message indicates to the preceding side that a connection has been established from this interface to the called party. On receipt of the CONNECT message, the preceding side enters the active state.

Under normal conditions, call clearing is initiated at the UNI. However, call clearing may also be initiated at a PNNI switch for network-related reasons. The clearing procedures are symmetrical and may be initiated by either the preceding or succeeding side of the PNNI. In the interest of clarity, the following procedures describe the case when the preceding side initiates clearing.

The preceding side initiates connection clearing by sending a RELEASE message. It disconnects the virtual channel (i.e., does not forward data cells of the connection being cleared) and enters the *release request* state. The succeeding side of the PNNI enters the *release indication* state upon receipt of a RELEASE message. This message then prompts the succeeding side to release the virtual channel, and to initiate procedures for clearing the connection by informing the PNNI call control entity. Once the virtual channel used for the call has been released, the succeeding side sends a RELEASE COMPLETE message to the preceding side, releases both the call reference and virtual channel (i.e. connection identifier and resources associated with the connection), and enters the null state.

Similarly, upon receiving the RELEASE COMPLETE message, the preceding side releases the virtual channel (i.e. connection identifier and resources associated with the connection), releases the call reference, and enters the null state.

7.3.2   A Point-to-Multipoint Call Setup

Point-to-multipoint call setup procedures in PNNI depend on whether or not the connection to one or more leaves is already established across the interface. If the connection already exists, it is in an active state and new leaves are added using the point-to-multipoint substates and messages. If there is no connection established to at least one leaf across a PNNI interface, then the point-to-point signaling procedures are used. The SETUP message contains the *endpoint reference* IE and an indication that it is a point-to-multipoint call in the user plane connection configuration field. Let us consider Figure 7.9 and assume that a point-to-point connection is established (with a point-to-multipoint indication) from R1 to L1 across S1 and S2. R1 then sends a message to add two parties, L2 and L3. Since the connection is in an active state across R1 and S1, R1 sends an ADD PARTY message for L2 and L3. Since the connection is in the active state across S1 and S2, S1 sends an ADD PARTY message towards L2. On the other hand, the connection is not active (i.e., not established) across S1 and S3. Hence, S1 sends a SETUP message to S3 towards L3.

Hence, if the connection to at least one leaf has already been established across the interface in consideration, the succeeding side of the switch sends to its preceding side an add party indicator. Upon receiving this indicator, the preceding side sends a PNNI ADD PARTY message, enters the add party initiated substate, and starts awaiting the ADD PARTY ACKNOWLEDGE message. The ADD PARTY message has all the information included in the original SETUP message used to set up the first connection (i.e., traffic parameters, service requirements, called party address, etc.).

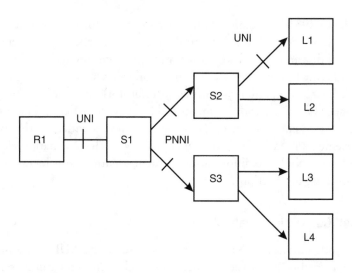

Figure 7.9 Point-to-multipoint connection setup at a PNNI.

Upon receiving an ADD PARTY message, the succeeding side of the next switch along the path enters the add party received substate for that leaf. Upon receiving an indication that the add party request has been accepted by the leaf, the succeeding side sends an ADD PARTY ACKNOWLEDGE message and enters the active substate. When the preceding side receives this acknowledgment, it enters the active substate.

Call clearing may be initiated at the UNI either by the root to drop a party or by a leaf to drop itself from the connection. It may also be initiated at a PNNI for network-related reasons. The clearing procedures are symmetrical and may be initiated by either the preceding or succeeding side of the PNNI. In the interest of clarity, the following procedures describe the case when the preceding side initiates clearing.

The preceding side initiates clearing of a leaf by sending a DROP PARTY message and enters the drop party initiated substate for that leaf. The succeeding side enters the drop party received state upon receipt of this message. The succeeding side then sends to the preceding side a DROP PARTY ACKNOWLEDGE message to indicate that the DROP PARTY message is received. Resources allocated to a call/connection are released at a PNNI only after the last leaf of the point-to-multipoint connection at that interface is cleared.

7.4 PNNI Point-to-Point Call/Connection Control Procedures

PNNI messages are used to manage user connections at a PNNI. There are also messages used to manage the interface itself. The PNNI message format is identical to the Q.2931 UNI signaling message format defined in Section 1.4. The PNNI and UNI messages are distinguished by their corresponding identifiers at the protocol discriminator field of each message. Accordingly, each PNNI message includes the PNNI protocol discriminator, call reference, message type (including message compatibility instruction indicator), and the message length. Message-specific information elements follow these four fields. The format of information elements used in PNNI signaling is also identical to that of information elements used in Q.2931 UNI signaling. Each information element is identified uniquely by its information element identifier and includes various fields as discussed in Section 1.5.

7.4.1 Call/Connection Establishment

A PNNI node initiates a PNNI connection by sending a SETUP message that causes the preceding side to start timer T303 and to enter the call present state. The message contains a call reference used to identify the connection at the interface in addition to all the information required for the succeeding side to process the call, such as service category, traffic parameters, and so forth.

7.4.1.1 No Reply to the First SETUP Message

If a node determines that the requested service category is not available, it initiates crankback with a cause. Similarly, if the *broadband bearer capability, ATM traffic descriptor, end-to-end transit delay*, and *extended QoS parameters* IEs contain a unsupported set of parameters, the succeeding side returns a RELEASE COMPLETE message with a cause "unsupported combination of traffic parameters" that also includes a *crankback* IE.

If no response to the SETUP message is received by the preceding side before the first expiration of timer T303, then the SETUP message may be retransmitted and timer T303 is restarted. If the preceding side does not receive any response to the SETUP until the timer expires for the second time, it enters the null state, sends a RELEASE COMPLETE message to the succeeding side with cause "recovery on timer expiry," and initiates clearing (without crankback) towards the calling party with cause "recovery on timer expiry." This sequence of events is illustrated in Figure 7.10.

7.4.1.2 Call Termination After the CALL PROCEEDING Message Is Sent

Upon receiving the SETUP message, the succeeding side enters the call initiated state and sends a CALL PROCEEDING message to the preceding side. This reply acknowledges the SETUP message, indicates that the call is being processed, and passes the connection identifier to the preceding side. The succeeding side then enters the call proceeding sent state. When the preceding side receives

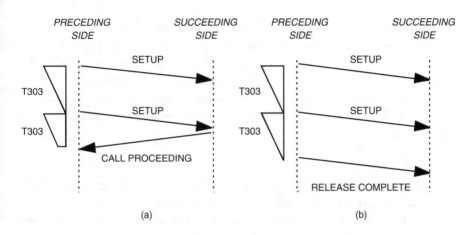

Figure 7.10 Timer T303 processing: (a) first and (b) second expiry of timer T303.

the CALL PROCEEDING message, it stops timer T303, starts timer T310, and enters the call proceeding received state. If, following the receipt of a SETUP message, the succeeding side determines that for some reason the call cannot be supported, it initiates call-clearing procedures. Similarly, if the preceding side has received a CALL PROCEEDING message, but does not receive a CONNECT, ALERTING, or RELEASE message prior to the expiration of the timer T310, it initiates clearing procedures (without crankback) towards the called party with cause "recovery on timer expiry." This sequence of events is illustrated in Figure 7.11.

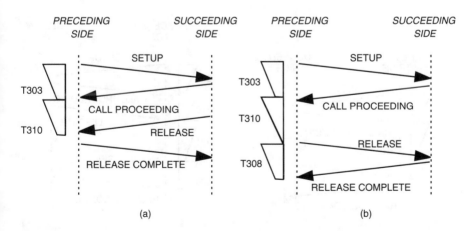

Figure 7.11 Timer T310 processing: (a) call clearing after receiving CALL PROCEEDING message and (b) timer T310 expiry.

7.4.1.3 Call Clearance After the ALERTING Message Is Sent

Upon receiving an indication that the called party is alerted, the succeeding side sends an ALERTING message to the preceding side and enters the alerting delivered state. When the preceding side receives an ALERTING message from the succeeding side, it stops timer T310, starts timer T301, enters the alerting received state, and sends an alerting indication towards the calling user (see Figure 7.12).

7.4.1.4 Call Clearance

Call clearing is normally initiated at the UNI and progresses across the PNNI network between the called and calling parties. However, it may also be initiated by the PNNI due to a failure, administrative action, or other exception condition.

The clearing procedures at the PNNI are symmetrical and may be initiated by either the preceding or the succeeding side. The following procedure applies to the case when the preceding side initiates the clearing (see Figure 7.13).

The preceding side initiates clearing by sending a RELEASE message, starting timer T308, disconnecting the virtual channel, and entering the release request state. Upon receipt of a RELEASE message, the succeeding side of the PNNI interface enters the release indication state and initiates procedures for clearing the network connection by informing the PNNI call control entity. Once the virtual channel used for the call has been released, the succeeding side sends a RELEASE COMPLETE message to the preceding side, releases both the call reference and the virtual channel (i.e., connection identifier), and enters

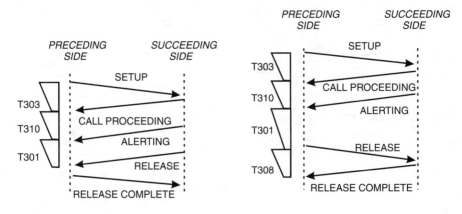

Figure 7.12 Timer T301 processing.

the null state. The RELEASE COMPLETE message has only local significance and does not imply an acknowledgment of an end-to-end clearing.

Upon receiving the RELEASE COMPLETE message, the preceding side stops timer T308, releases the virtual channel, releases the call reference, and returns to the null state. If timer T308 expires for the first time, the preceding side retransmits a RELEASE message to the succeeding side with the cause number originally contained in the first RELEASE message, restarts timer T308, and remains in the release request state. In addition, the preceding side may indicate a second cause IE with cause "recovery on timer expiry." If no RELEASE COMPLETE message is received from the succeeding side before timer T308 expires the second time, the preceding side releases the call reference and returns to the null state.

In general, a message requesting a certain action may be sent by the both sides of the PNNI almost simultaneously. Clear collision occurs when both sides simultaneously transfer a RELEASE message related to the same call reference value. If the preceding or succeeding side receives a RELEASE message while in the release request state, the receiving entity stops timer T308, releases the call reference and virtual channel, and enters the null state (without sending or receiving a RELEASE COMPLETE message). Similarly, collision can occur when both sides of the PNNI interface simultaneously transfer SETUP messages with traffic parameters so that it is not possible to establish both connections simultaneously. Both sides of the PNNI interface in this case are required to clear the call/connection with a crankback IE with one of the following causes: "resources unavailable," "unspecified," "quality of service unavailable," or "user cell rate not available."

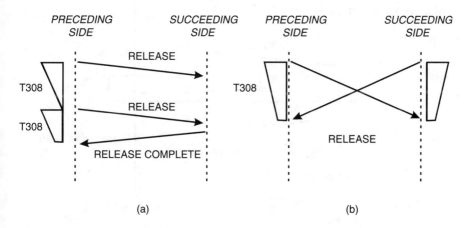

Figure 7.13 Call clearing: (a) timer T308 expiry and (b) clear collision.

7.4.1.5 Successful Call Establishment and Termination

The preceding side sends a SETUP message to request a connection establishment between itself and the neighbor switch based on the path information included in the *DTL* IE of the SETUP message. The succeeding side sends a CALL PROCEEDING message to acknowledge the receipt of the SETUP message. Upon receiving an indication that the called party is alerting, the succeeding side sends an ALERTING message to indicate that the called party is being alerted.

Upon receiving an indication from call control that the call has been accepted, the succeeding side sends a CONNECT message to the preceding side and enters the active state. This message indicates to the preceding side that a connection has been established from this interface to the called party. On receipt of the CONNECT message, the preceding side stops timer T310 (if an ALERTING message was not sent/received) or timer T301 (if ALERTING message is received) and enters the active state. Figure 7.14 illustrates a successful connection setup.

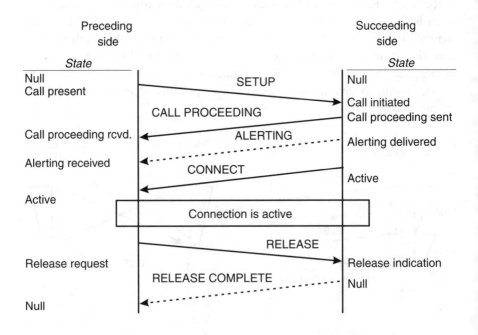

Figure 7.14 A successful call/connection setup.

7.4.2 PNNI Point-to-Point Call Control Messages

In this section, PNNI messages used to manage point-to-point call/connection control are reviewed in detail. Each PNNI signaling message contains one or more information elements as needed, describing various aspects of the interaction as defined by the message. Various information elements used in PNNI signaling are the same as in UNI signaling as described in Chapter 2. In this section, we discuss only the PNNI-specific parts of these messages.

7.4.2.1 ALERTING

The ALERTING message at the PNNI indicates that the called user has been alerted for the incoming call. This message, initiated by the called user, travels across a PNNI network from the called user towards the calling user.

The endpoint reference IE is used in this message only for point-to-multipoint connections to identify the end point corresponding to the called user. Its inclusion is mandatory if an endpoint reference was included in the setup indication. A *notification indicator* IE is a general purpose element used to carry information either end-to-end between end users or between the network and the calling user. For example, if the call goes through an interworking unit (e.g., ATM to ISDN or ATM to frame relay), this information element will inform the calling user that interworking is taking place for this call. It is included if the received alerting indication contains this information (see Figure 7.15).

7.4.2.2 CALL PROCEEDING

The CALL PROCEEDING message at the PNNI is used to acknowledge the receipt of a SETUP message. Before the succeeding node that received the SETUP message passes the SETUP message to the next PNNI node along the connection's path (as specified in the DTLs), it transmits the CALL PROCEEDING message to its preceding node. In addition to acknowledging the receipt of the SETUP message, this message is also used to negotiate the connection identifier (i.e., VPI/VCI) between two PNNI nodes (See Figure 7.16).

The endpoint reference IE is mandatory if an endpoint reference was included in the setup indication.

Information Element	Type	Length
Endpoint reference	O	4–7
Notification indicator	O	4–*

Figure 7.15 Information elements used in ALERTING message.

Information Element	Type	Length
Connection identifier	M	9
Endpoint reference	O	4–7

Figure 7.16 Information elements used in CALL PROCEEDING message.

7.4.2.3 CONNECT

The CONNECT message is used to indicate call acceptance and that the end-to-end path is established from the called party to the switch that sends the CONNECT message. It travels from the called user towards the calling user along the end-to-end path.

PNNI nodes transport *broadband low-layer information, connected number, connected subaddress, called party soft PVPC/PVCC,* and *AAL parameters* IEs transparently. A called party soft PVPC/PVCC (permanent virtual path connection / permanent virtual channel connection) IE is used to specify the connection identifier at the UNI the called user is attached. This information element is discussed in detail in Section 7.5.6.

Various negotiated parameters are carried in the CONNECT message. In the case of ABR service, two information elements are used: ABR additional parameters and ABR setup parameters (the current version of the PNNI specification does not include ABR virtual source/virtual destination behavior support) are used (i.e., copied) to determine the final values of ABR parameters and allocate resources to the connection. The inclusion of the ATM traffic descriptor IE is mandatory if the calling user requested an ABR traffic category connection. The details of the ABR processing are discussed in Section 7.5.8.

ATM traffic descriptor, end-to-end transit delay, and extended QoS parameters in the CONNECT message are used at a PNNI node to adjust and allocate the amount of resources reserved during the time the SETUP message was proceeding from the calling user towards the called user. Their values are not changed in the CONNECT message (see Figure 7.17).

7.4.2.4 SETUP

The SETUP message is used to initiate a call setup at a PNNI. Accordingly, this message includes all the information required to establish a call at the interface. The mandatory information elements include the ATM traffic descriptor identifying the traffic parameters of the call, the broadband capability identifying the service class (e.g., traffic type) requested for the call, the called party number identifying the destination end station, and the designated transit list (DTL). DTL IE, discussed in detail in Section 7.5.4.1, identifies the end-to-end path as

Information Element	Type	Length
ABR additional parameters	O[1]	4–14
ABR setup parameters	O[1]	4–36
AAL parameters	O[1].	4–11
ATM traffic descriptor	O	4–30
Broadband low-layer information	O[1]	4–17
Called party soft PVPC/PVCC	O	4–11
Connected number	O[1]	4–26
Connected subaddress	O[1]	4–25
Endpoint reference	O	4–7
End-to-end transit delay	O[1]	4–7
Extended QoS parameters	O[1]	4–13
Generic identifier transport	O[1]	4–33
Notification indicator	O[1]	4–*

[1]Included if the received connect indication contains this information.

Figure 7.17 Information elements used in CONNECT message.

chosen by the originating switching system. The broadband repeat indicator is another mandatory IE used to indicate the order of the DTL IEs in the DTL stack. It is present even when there is only one designated transit list IE.

In addition to these mandatory IEs, there are optional information elements that may be included in the SETUP message. Some of these optional elements are transported across a PNNI transparently. These include AAL parameters, broadband high-layer information, broadband low-layer information, called party subaddress, calling party subaddress, calling party number, generic identifier transport, and notification indicator (see Figure 7.18).

Soft PVPC- or PVCC-related information elements are included in the SETUP message only when a soft PVPC or PVCC is being established across a PNNI. These information elements are also transparent to PNNI nodes and have significance only at the originating and terminating PNNI nodes that are attached to the corresponding UNI segments. This capability is explained in detail in Section 7.5.6. Calling party soft PVPC/PVCC IE is included in case of soft PVPC/PVCC setup, when the calling endpoint wants to inform the destination network interface of the values used for PVC segment at the calling end.

A connection identifier IE is included when the preceding side wants to indicate a particular virtual channel. If not included, its absence is interpreted as

Information Element	Type	Length
AAL parameters	O[1]	4–21
ABR additional parameters	O	4–14
ABR setup parameters	O	4–36
Alternative ATM traffic descriptor	O	4–30
ATM traffic descriptor	M	12–30
Broadband bearer capability	M	6–7
Broadband high-layer information	O[1]	4–13
Broadband repeat indicator	O[1]	4–5
Broadband low-layer information	O[1]	4–17
Called party number	M	4-25
Called party soft PVPC/PVCC	O[1]	4–11
Called party subaddress	O[1]	4–25
Calling party number	O[1]	4–26
Calling party soft PVPC/PVCC	O	4–10
Calling party subaddress	O[1]	4–25
Connection identifier	O	4–9
Connection scope selection	O[1]	4–6
Broadband repeat indicator	M	5
Designated transit list	M	33–546
Endpoint reference	O[1]	4–7
End-to-end transit delay	O	4–13
Extended QoS parameters	O	4–25
Generic identifier transport	O[1]	4–33
Minimum acceptable ATM traffic descriptor	O[1]	4–20
Notification indicator	O[1]	4–*
QoS parameter	O[1]	4–6
Transit network selection	O[1]	4–8

[1]Included if the received setup indication contains this information.

Figure 7.18 Information elements used in SETUP message.

meaning that any virtual channel is acceptable. The assignment of virtual channels across a PNNI is discussed in detail in Section 7.5.1.

If the requested broadband bearer capability is the ABR service, then the inclusion of the ABR additional parameters and ABR setup parameters is mandatory in the SETUP message. The details of ABR-related information elements are discussed in Section 7.5.8.

If a network needs to discard cells in order to avoid getting into a congested state or to recover from such a state, it may be desirable to discard more than one consecutive cell on a given connection. This is especially true for connections supporting applications where the information is organized into frames consisting of two or more ATM cells. The frame discard option during the call establishment allows the network nodes to detect the frame boundary by examining the payload type field of the ATM cell header and to discard frames from these connections as needed to relieve the congestion.

PNNI allows the user and the network to negotiate traffic parameters during the call setup time. The negotiation is achieved by including either the minimum acceptable ATM traffic descriptor or the alternative ATM traffic descriptor IE. The minimum acceptable ATM traffic descriptor IE indicates the minimum value of each traffic parameter included in the ATM traffic descriptor IE that may be acceptable for the call. The alternative ATM traffic descriptor IE may be used to indicate a set of values different to choose from than the ones included in the ATM traffic descriptor IE for the switching systems.

The agreed upon values (as determined together by the PNNI network and the destination end user) are carried from the called end user to the calling end user in the PNNI CONNECT message within the PNNI network in the ATM traffic descriptor IE and UNI CONNECT message at the UNI in the ATM traffic descriptor IE. The details of the negotiation process are described in Section 7.5.2.

The characterization of the call also includes the service requested from the network. The QoS request for the call is specified in QoS parameter, end-to-end transit delay, and extended QoS parameters IEs. The latter two IEs are used to specify individual QoS parameters. In particular, the extended QoS IE includes CDV and CLR, whereas the end-to-end transit delay IE is used for CTD. The QoS parameter IE is used to specify the QoS class requested for the call. This information element is used for backward compatibility with UNI 3.1 and for interworking with networks that support ITU recommendations.

The QoS parameter IE is required in the UNI 3.1 SETUP message, whereas it is optional in UNI 4.0. It is included in the UNI 4.0 SETUP message if individual parameters are not specified. When included in the UNI 4.0 SETUP message, the QoS parameter IE is mandatory in the PNNI SETUP message as well. However, a PNNI node may transport this information element transparently. The inclusion of end-to-end transit delay and extended QoS parameter IEs depends on the ATM transfer capability specified in the bearer capability IE. The details of QoS parameter selection procedures are described in Section 7.5.3.

A *connection scope* IE is used to indicate the scope of the anycast address.

A *transit network selection* IE is used to select a non-PNNI network used to provide connectivity from a PNNI node to the called party. The called party could be attached to the transit network or there may be a path across the transit network to the network the called party is attached to. The interface between a PNNI node and the transit network is a UNI.

7.5 Additional PNNI Signaling Procedures

In this section, we review various PNNI signaling procedures used to manage switched connections across a PNNI.

7.5.1 Connection Identifier Allocation

As a connection is set up, a connection identifier is allocated to it across each interface. This requires the preceding and succeeding nodes to agree upon the value of the connection identifier used for each connection.

The ATM cell header has a 12-bit VPI and a 16-bit VCI connection identifier field. In PNNI, the range of valid VPI values is 0 to 4,095. However, some values in this range may not be available to use for switched VPs (e.g., some values may be used for permanent virtual path connections). A switching system may also restrict the upper bound of the VPI range by restricting the number of active VPI bits to less than 12 bits.

Each VPI can be used to define up to 2^{16} connections by distinguishing among different connections using the VCI field. However, some of the VCI values, independent of the VPI value, are reserved as follows:

- 0–31—Not used for on-demand user plane connections;
- 32–65,535—May be used for on-demand user plane connections.

A switching system may impose a limit on the VCI range by restricting the number of VCI bits (to less than 16) available for on-demand user connections. Furthermore, some of the values in the range 32 to 65,535 may not be available for use by on-demand user plane connections.

It is possible that both sides of a PNNI may transfer a SETUP message almost simultaneously. If both sides indicate the same connection identifier (VPI for a switched virtual path and VPI/VCI for a switched virtual channel), then a call collision occurs. As a general rule, the side that has the higher node identifier has the authority to allocate the connection identifier value to avoid call collision. Additional procedural issues are discussed next in the context of switched virtual channels and switched virtual paths.

7.5.1.1 Switched Virtual Channels

There are two types of signaling channels used in ATM, associated signaling and nonassociated signaling, as discussed in Section 1.2.

Associated Signaling

In associated signaling, the preceding side requests a VC in a VPC to carry signaling messages for that VP. This requires the VP associated signaling field of the *connection identifier* IE to be coded as "VP associated signaling." Furthermore, the preceding side specifies in the preferred/exclusive field either "exclusive VPI; any VCI" or "exclusive VPI; exclusive VCI."

Exclusive VPI; Any VCI. The succeeding side selects any VCI available within the VPC. If no VCI is available, a RELEASE COMPLETE message that includes a crankback IE (with cause "no VPI/VCI available") is sent by the succeeding side.

Exclusive VPI; Exclusive VCI. The preceding side requests a specific VCI in a VP. If the indicated VCI is available, the succeeding side selects it for the call. If it is not available, a RELEASE COMPLETE message that includes a crankback IE (with the cause "requested VPI/VCI not available") is sent by the succeeding side. Table 7.5 lists the use of connection identifier options in the SETUP and CALL PROCEEDING messages with associated signaling.

Table 7.5
Connection ID Options in PNNI Signaling With Associated Signaling

Preceding Node	Connection ID Options in SETUP	Connection ID Options in CALL PROCEEDING
Higher node identifier	VP associated signaling; exclusive VPI; exclusive VCI	VP associated signaling; exclusive VPI; exclusive VCI
Lower node identifier	VP associated signaling; exclusive VPI; any VCI	VP associated signaling; exclusive VPI; exclusive VCI

In both cases, the selected VCI value is indicated in the connection identifier IE in the CALL PROCEEDING message.

Nonassociated Signaling

When the preceding side requests a virtual channel in the SETUP message, it indicates one of the following in the VP associated signaling field of the connection identifier IE.

Exclusive VPI; Any VCI.   If the indicated VPI and a VCI in the chosen VPI are both available, the succeeding side selects one VCI for the call. If the specified VPI or a VCI in the requested VPI is not available, a RELEASE COMPLETE message that includes a crankback IE (with cause "requested VPI/VCI not available") is sent by the succeeding side.

Exclusive VPI; Exclusive VCI.   If both the indicated VPI and the indicated VCI are available, the succeeding side selects the identifiers for the call. If either the specified VPI or the VCI is not available, a RELEASE COMPLETE message that includes a crankback IE (with cause "requested VPI/VCI not available") is sent by the succeeding side.

No Indication (The Connection Identifier IE Is Not Included in the SETUP Message).   The succeeding side selects any available VPI and VCI. If the succeeding side is not able to allocate a VCI in any VPI, a RELEASE COMPLETE message that includes a crankback IE (with cause "no VPI/VCI available") is sent by the succeeding side.

The selected VPI/VCI is indicated in the connection identifier IE in the first message returned by the succeeding side in response to the SETUP message (i.e., CALL PROCEEDING). The VP associated signaling field in the CALL PROCEEDING message is coded as "explicit indication of VPI" and the preferred/exclusive field is coded as "exclusive VPI; exclusive VCI."

In order to avoid call collision, the side that has the higher node identifier allocates the connection identifier values. The side with the higher node identifier uses the "exclusive VPI and exclusive VCI" in the connection identifier IE of the SETUP message. A SETUP message generated from the side with the lower node identifier can use the other two options. Table 7.6 lists the use of connection identifier options in the SETUP and CALL PROCEEDING messages with nonassociated signaling.

7.5.1.2 Switched Virtual Paths

To request the establishment of a SVP, the bearer class field of the broadband bearer capability IE in the SETUP message is set to indicate "VP service." The preceding side indicates one of the following:

Table 7.6
Connection ID Options in PNNI Signaling With Nonassociated Signaling

Preceding Node	Connection ID Options in SETUP	Connection ID Options in CALL PROCEEDING
Higher node identifier	Explicit indication of VPI; exclusive VPI; exclusive VCI	Explicit indication of VPI; exclusive VPI; exclusive VCI
Lower node identifier	Explicit indication of VPI; exclusive VPI; any VCI	Explicit indication of VPI; exclusive VPI; exclusive VCI
	No connection ID information element	Explicit indication of VPI; exclusive VPI; exclusive VCI

- *No indication is included* (i.e., the connection identifier IE is not included in the SETUP message)—The succeeding side selects any available VPI. If the succeeding side is not able to allocate a VPI, a RELEASE COMPLETE message that includes the crankback IE (with cause "no VPI/VCI available") is sent by the succeeding side.
- *Exclusive VPI and no VCI*—If the indicated VPI is available, the succeeding side selects it for the call. If the indicated VPI is not available, a RELEASE COMPLETE message that includes a crankback IE (with cause "requested VPI/VCI not available") is sent by the succeeding side.

The selected VPI is indicated in the connection identifier IE in the first message returned by the succeeding side in response to the SETUP message (i.e., CALL PROCEEDING). The VP associated signaling field is coded as "explicit indication of VPI" with the preferred/exclusive field coded as "exclusive VPI; no VCI."

Similar to the case of switched SVCs, call collision may occur when both sides of an interface simultaneously transfer SETUP message indicating the same exclusive VPI. The side with the higher node identifier may use the option "exclusive VPI and no VCI" in the connection identifier IE of the SETUP message. A SETUP message from a node with a lower node identifier uses the other option. Table 7.7 lists the connection identifier options for switched virtual paths.

Table 7.7
Connection ID Options for Switched Virtual Paths

Preceding Node	Connection ID Options in SETUP	Connection ID Options in CALL PROCEEDING
Higher node identifier	Explicit indication of VPI; exclusive VPI; no VCI	Explicit indication of VPI; exclusive VPI; no VCI
Lower node identifier	No connection ID information element	Explicit indication of VPI; exclusive VPI; no VCI

7.5.2 Traffic Parameter Negotiation During Call/Connection Setup

A SETUP message includes a traffic descriptor IE to specify the call/connection traffic characteristics to the switching system. A switching system uses the traffic parameters and QoS requirement to make a call admission decision by taking into consideration the connections already established. An ATM end station may be capable of sending its traffic to the network with different sets of traffic parameters by shaping its traffic before it is submitted to the network. For example, instead of sending traffic at the interface rate, the peak rate of the connection may be shaped at the end station to a rate lower than that of the interface. This might in turn increase the chance of a call/connection request being accepted by the switching system.

PNNI negotiation procedures allow each switching system to adjust the parameters of a call as needed. These procedures require the user to initiate the negotiation by including in the SETUP message either the minimum acceptable ATM traffic descriptor IE or the alternative ATM traffic descriptor IE (but not both). If both information elements are included in the SETUP message, the call request is rejected without crankback (with cause "unsupported combination of traffic parameters").

If the alternative ATM traffic descriptor IE is used, the switching system uses only the two sets of values defined in the ATM traffic descriptor and alternative ATM traffic descriptor IEs in its internal call admission procedures. If the switching system can meet the service requirements of the call/connection with the either one of the two sets of values, it may accept the call; otherwise, it rejects the call.

If a node is able to provide both traffic parameter values specified in the ATM traffic descriptor and alternative ATM traffic descriptor IEs, it progresses the connection-establishment request with both information elements.

If a node is able to provide the traffic parameter values specified in the ATM traffic descriptor IE, but not the values specified in the alternative ATM traffic descriptor IE, it progresses the connection-establishment request with the ATM traffic descriptor IE only.

If a node is not able to provide the traffic parameter values specified in the ATM traffic descriptor IE, but it is able to provide the traffic parameter values specified in the alternative ATM traffic descriptor IE, it progresses the connection-establishment request by using the contents of the alternative ATM traffic descriptor IE as the ATM traffic descriptor. In this case, the alternative ATM traffic descriptor IE is not forwarded.

If a node cannot support either set of traffic parameter values, it rejects the connection-establishment request with the cause "user cell rate unavailable." The RELEASE message includes a crankback IE with the corresponding crankback cause code (see Figure 7.19).

Unlike the alternative traffic descriptor IE, the minimum acceptable ATM traffic descriptor IE allows the end station to specify a range of values for each parameter: one set of values in the ATM traffic descriptor and another in the minimum acceptable ATM descriptor. The cell rates in the minimum

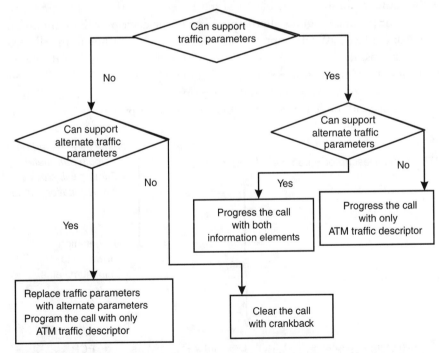

Figure 7.19 Traffic parameter negotiation with the alternate traffic parameters.

acceptable traffic descriptor IE are required to be less than their corresponding values defined in the ATM traffic descriptor IE.

If a node is able to provide the traffic parameter values specified in the ATM traffic descriptor IE, it progresses the connection-establishment request with both the ATM traffic descriptor IE and the minimum acceptable ATM traffic descriptor IE. If it is not able to provide some of the cell rates indicated in the ATM traffic descriptor IE, but it is able to provide at least their corresponding cell rates in the minimum acceptable ATM traffic descriptor IE, it progresses the connection-establishment request after adjusting the cell rates in the ATM traffic descriptor IE with reduced values at or above those specified minimum acceptable values (see Figure 7.20).

If a node progresses the connection request to the next node along its path after adjusting the cell rates, it checks whether or not some of the cell rates are still above the minimum rates defined in the minimum acceptable ATM traffic descriptor IE. If some of the adjusted parameters are still greater than their minimum values, the call is progressed with both information elements, containing only parameters that may be adjusted at the downstream nodes. Otherwise, the call progresses with the modified ATM traffic descriptor IE and without the minimum acceptable ATM traffic descriptor IE.

If the parameters of the alternative ATM traffic descriptor IE or minimum acceptable ATM traffic descriptor IE are not specified according to the allowed combinations, the network handles these information elements as if they were nonmandatory information elements with content errors.

If a peer returns the ATM traffic descriptor IE in the CONNECT message, the network node forwards the ATM traffic descriptor IE as it received it.

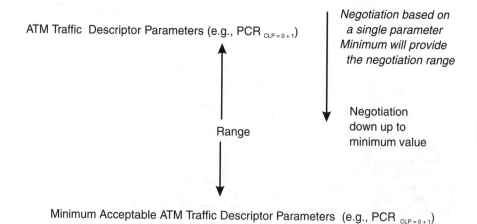

Figure 7.20 Negotiation of the traffic parameters.

If the traffic parameter values in the CONNECT message differ from those that the node has forwarded in the SETUP message, the resources that the node has allocated to the connection are modified accordingly.

If a peer returns no ATM traffic descriptor IE in the CONNECT message and the network node has negotiated the traffic parameters when it processed the SETUP message, the node includes the ATM traffic descriptor IE in the CONNECT message before forwarding it.

If a peer returns no ATM traffic descriptor in the CONNECT message and the network node has not negotiated the traffic parameters when it processed the corresponding SETUP message, the node can optionally include the ATM traffic descriptor IE in the CONNECT message.

7.5.3   QoS Parameter Selection Procedures

QoS requirements are expressed in PNNI networks using individual quality of service parameters and/or the end-to-end transit delay IE. The individual QoS parameters are included in the extended QoS parameters IE.

The allowed set of individual QoS parameters in the SETUP message is determined by the ATM service category of the call. For each individual QoS parameter, if the end-to-end value of that parameter is determined by accumulation, the corresponding cumulative value of the parameter is included in the SETUP message. The cumulative forward and backward values of individual QoS parameters are updated sequentially along the route of the call.

When the preceding side receives a setup indication that includes a QoS parameter IE, it includes a QoS parameters IE in the corresponding SETUP message (see Figure 7.21).

At the UNI, it is mandatory to include the ATM service category of the call (i.e., CBR, RT-VBR, or NRT-VBR, UBR, or ABR). If a network node determines that the requested service category is not available, it initiates crankback with one of the following cause and crankback cause codes: "bearer capability not authorized," "bearer capability not presently available," or "bearer service not implemented."

It is not, however, required to include the extended QoS parameters IE. If the received setup request does not contain the extended QoS parameters and the previous interface is not a PNNI, the preceding side generates the extended QoS parameters based on the service category, using the default values defined for the service. That is, upon receiving a SETUP request without the extended QoS parameters IE, the preceding side of the PNNI generates an extended QoS parameters IE using a local mapping from the service category and the forward and backward QoS class fields in the QoS parameter IE. An end-to-end transit delay IE may also be generated as part of the above mapping (if it was not con-

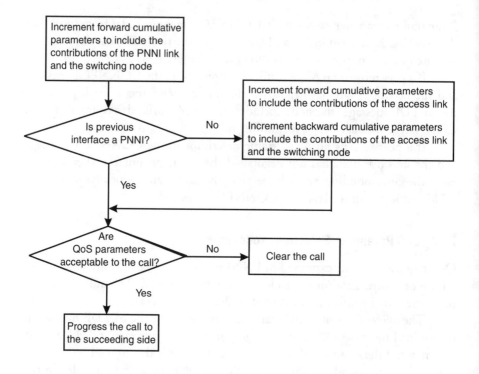

Figure 7.21 Preceding side QoS parameter processing.

tained in the received setup indication). When such a mapping is used, the node that generated the individual QoS parameters is marked as an intermediate network. This is achieved by setting the origin field in the extended QoS parameters IE to "intermediate network" and including a network generated indicator in the end-to-end transit delay IE. All cumulative parameter values generated from this mapping are initialized to 0 before starting to process the individual QoS parameters.

If the succeeding side detects that the broadband bearer capability, ATM traffic descriptor, end-to-end transit delay, and extended QoS parameters IEs contain a set of parameters, the succeeding side returns a RELEASE COMPLETE message with cause "unsupported combination of traffic parameters."

For each parameter contained in the extended QoS parameters IE and/or the end-to-end transit delay IE, the preceding side takes the following actions (whether the information element was contained in the received SETUP message or whether the information element was generated using a local mapping from the QoS class).

If the previous interface is not a PNNI interface and the parameter is cumulative, the preceding side increments the corresponding cumulative forward value and cumulative backward value to account for expected increases due to user data transfer over the previous link and backward data transfer within this switching system. After this operation, the cumulative value is equal to the accumulated value from the network boundary to this switching system.

The preceding side also determines whether or not acceptable values for each parameter can be supported. If a value cannot be supported, the preceding side follows the crankback procedures with cause "quality of service unavailable." If it is able to provide the acceptable values of all specified individual QoS parameters, the call is progressed to the succeeding side.

If no value for an individual QoS parameter for the corresponding ATM service category is specified in the extended QoS parameters or in the end-to-end transit delay IE, any value of the individual QoS parameter is acceptable (as a default) and the preceding side shall continue to process the call.

The succeeding side does not make use of the QoS parameter IE, but passes it on if it is present and if the call is progressed. For each parameter contained in the extended QoS parameters IE and/or the end-to-end transit delay IE, the succeeding side does the following (see Figure 7.22):

1. Increments the backward cumulative value of the parameter, if the parameter is cumulative, to account for the expected increases due to user data transfer within this switching system and over the link from this switching system to the switching system at the preceding side;

2. If the next interface over which the call is to be routed is not a PNNI interface and the parameter is cumulative, the succeeding network node shall increment the cumulative forward value of that parameter to account for the expected increases due to user data transfer within this switching system and over the following link between this switching system and the network boundary. In addition, it increments the cumulative backward value to account for the expected increases due to user data transfer over the following link from the network boundary to this switching system;

3. Determines if the highest/lowest acceptable values of that parameter can be supported. If no values {less than/greater than} or equal to the highest/lowest acceptable value can be supported, the succeeding side shall follow the crankback procedures by returning a RELEASE or RELEASE COMPLETE message (depending on whether or not a CALL PROCEEDING message has been sent yet) with cause and, if applicable, crankback cause "quality of service unavailable."

If no acceptable forward value of an allowed individual QoS parameter for the corresponding ATM service category is specified (in the extended QoS parameters or end-to-end transit delay IEs), the default is that any value of the individual QoS parameter is acceptable and the succeeding side continues to process the call.

7.5.3.1 End-to-End Transit Delay

The end user may specify its end-to-end transit delay requirement in the forward maximum cell transfer delay acceptable field of the end-to-end transit delay IE. The originating switch uses this value to find a path in the network (together with other traffic- and QoS-related information) that is likely to meet the service requirement of the connection. Once the end-to-end connection is set up in the network, the actual end-to-end delay (including the delays at the end stations) is required to be less than or equal to the requested value. Since the actual delay is not known exactly until the connection is set up in the network and accepted by each node along the path, there is a need to convey the actual delay to the calling end station in the CONNECT message. For this purpose, a

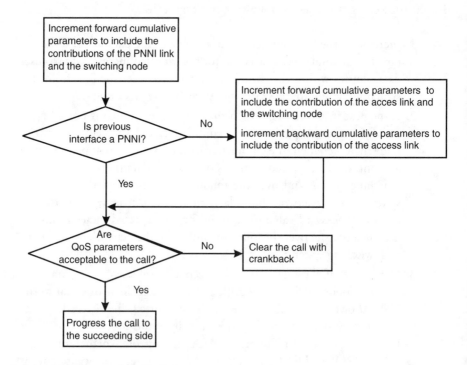

Figure 7.22 Succeeding side QoS parameter processing.

cumulative forward maximum cell transfer delay field is added to the end-to-end transit delay IE.

As the SETUP message progresses from the calling end station to the called end station, each node along the path updates the cumulative forward maximum cell transfer delay field by adding its own contribution to the end-to-end delay to the value it received from the preceding node. This additional delay includes all nodal delays (such as queuing delay, switching delay, etc.) and link delays (such as link transmission time and propagation delay).

The end-to-end transit delay IE illustrated in Figure 7.23 may be generated either by the calling user (at its UNI SETUP message) or by the PNNI network if it was not generated by the user. The network-generated indicator field is included in this information element when it is generated by the network. The UNI CONNECT message does not include this information element if it is generated by the network.

Octet groups 5 and 7 are mutually exclusive. Octet group 5 is used in the CONNECT and ADD PARTY ACKNOWLEDGE messages, and octet group 7 is used in the SETUP and ADD PARTY messages.

The acceptable forward maximum cell transfer delay and the corresponding cumulative value is expressed in the PNNI in units of microseconds. The

8	7	6	5	4	3	2	1	Octet
End-to end transit delay information element identifier								1
1 ext	1 Coding standard	1	IE Instruction field					2
Length of the End-to-end transit delay contents								3
								4
Cumulative Forward Maximum Cell Transfer Delay Identifier								5*
Cumulative Forward Maximum Cell Transfer Delay								5.1*
								5.2*
PNNI Acceptable Forward Maximum Cell Transfer Delay Identifier								6*
								6.1*
PNNI Acceptable Forward Maximum Cell Transfer Delay								6.2*
								6.3*
PNNI Cumulative Forward Maximum Cell Transfer Delay Identifier								7*
								7.1*
PNNI Cumulative Forward Maximum Cell Transfer Delay								7.2*
								7.3*
Network-generated indicator								8*

Figure 7.23 End-to-end transit delay IE.

value 1111 1111 1111 1111 1111 1111 is reserved to indicate that any forward maximum cell transfer delay value is acceptable. The cumulative forward maximum cell transfer delay in a UNI (if included), on the other hand, is expressed in units of milliseconds. Although this granularity is sufficient for end stations, this granularity is rather large inside the PNNI network. As an example, consider a PNNI call traversing one or more local area networks with delays in each LAN being on the order of tens of microseconds. If a millisecond granularity is used in accumulating the end-to-end delays, a large number of calls may be rejected artificially.

7.5.4 Designated Transit List

The end-to-end path of a connection in PNNI is determined by the switch the connection originates at its UNI interface (i.e., originating switch). A sequence of logical nodes and possibly logical links that specify the chosen end-to-end path is placed in the DTL IE included in both the SETUP and ADD PARTY messages. These information elements are processed at each node along the path to determine the address of the next (logical) node along the end-to-end path.

7.5.4.1 DTL Information Element

Figure 7.24 illustrates the format of the DTL IE. The current transit pointer is an offset pointer to the transit node/logical port that indicates the current call/connection's progress along the DTL. The value 1 indicates the first transit in the information element (i.e., the node that generated the information element), the value 27 indicates the second transit (i.e., the nearest neighbor of the node that generated the information element), the value 54 indicates the third transit, and so on.

The logical node identifier uniquely identifies the node (i.e., a real switching system or a peer group at some level of hierarchy) that the call/connection is to transit, whereas the logical port identifier uniquely identifies a port of the logical node that the call/connection is to transit. The combination of the node identifier and logical port identifier unambiguously (but not uniquely) identifies a logical link. A node identifier is 22 bytes long (octets 7.1 to 7.22), whereas a port identifier is 4 bytes long (octets 7.23 to 7.26). Specification of port IDs in DTLs is optional. If a port ID is set to 0, it indicates that any port to the next node in the DTL stack can be used.

Each node port pair is placed in sequence consecutively at the end of the information element. The octet group 7 (a total of 26 bytes) in a DTL IE may be repeated as many as 20 times. This allows the specification of a path with a maximum of 20 hops within a peer group. A SETUP or an ADD PARTY mes-

8	7	6	5	4	3	2	1	Octet
Designated transit list Information element identifier								1
1 ext	Coding standard	IE Instruction Field						2
Length of designated transit list contents								3–4
Current transit pointer								5–6
Logical node / logical port indicator								7
Logical node identifier								7.1 to 7.22
Logical port identifier								7.23 to 7.26

Figure 7.24 DTL information element.

sage may have up to 10 DTL IEs, allowing 10 logical groups to be crossed (possibly at different levels of the hierarchy).

One DTL IE is used to specify the sequence of nodes/ports along the path for a given peer group. For example, let us consider a connection that originates and terminates within one peer group. The end-to-end path crosses A.1, A.3, A.4, and A.6. In this case, the signaling message includes one DTL IE with four logical node/port pairs each, corresponding to a physical node along the end-to-end path. The first octet group is the address of A.1, followed by the second octet group corresponding to A.3, and so on. In this example, port identifiers are not specified (hence set to 0 in the IE).

Say the an end-to-end connection crosses three peer groups, A, C, and D, all at the same level of the hierarchy. The connection originates at peer group A and within A crosses the switches A.1, A.3, and A.5. In this case, the originating switch creates two DTL IEs: one to specify the sequence of nodes/ports within its own peer group (i.e., A.1, A.3, A.5) and another to specify the node/port identifiers of each peer group along the path (i.e., A, C, D).

Hence, each DTL IE specifies a path in a given hierarchy, and there are as many DTL IEs in the signaling message as the number of hierarchical levels along the end-to-end path from the originating switch's view of the network. The designated transit list is an ordered list. The interpretation of the octet groups within the DTL IE is position dependent.

When a node receives a PNNI SETUP or ADD PARTY message, the current transit pointer of the DTL IE should point to the node that received the message at some level of the hierarchy. Before passing the signaling message to the next node along the path, the current node advances the current transit pointer from itself to the next node in the list. Revisiting the connection setup example within a peer group with the DTL A.1, A.3, A.5, node A.1 processes the SETUP message and advances the pointer to the second node in the list (i.e., A.3). Similarly, node A.3 advances the pointer to A.5 before transmitting the signaling message, and so on until it reaches to the node that has direct connectivity to the destination end station.

In summary, a DTL indicates the logical nodes and possibly logical links (i.e., transits) that a connection is to traverse through a peer group at some level of hierarchy. The hierarchical representation of network topology leads to a hierarchically complete source route as a stack (last in, first out list) of DTLs. The stack of DTLs is represented in the SETUP and ADD PARTY messages as a sequence of DTL IEs, one information element for a complete path in a particular hierarchy.

Each DTL IE contains, in the order that they are to be traversed, a list of transit nodes at a single level of the hierarchy. This list starts with the first logical node to be traversed at that level and ends at the last logical node to be traversed at that level. Whenever the path changes from one level of hierarchy to another, a DTL IE is pushed onto the stack and the current transit pointer is set to 1, pointing to the first logical node (and optionally logical link) in the designated transit list.

7.5.4.2 DTL Processing

The DTL processing procedures carried out at each node along the path of a call consist of two parts:

- Determining whether any DTLs need to be added to the DTL stack and taking appropriate measures (Figure 7.25);
- Determining whether any DTLs need to be removed from the stack and processing the remaining DTL stack to prepare it for transmission to the next node (Figure 7.26).

Hence, a node either adds to the DTL stack or removes a DTL from it. Each PNNI node along the path of a call is categorized based on its role in DTL processing: (1) DTL originator, (2) entry border node, (3) neither DTL origina-

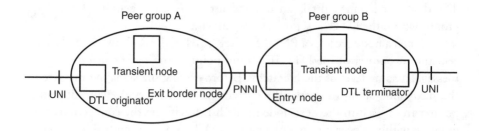

Figure 7.25 PNNI nodes classified based on their roles in DTL processing.

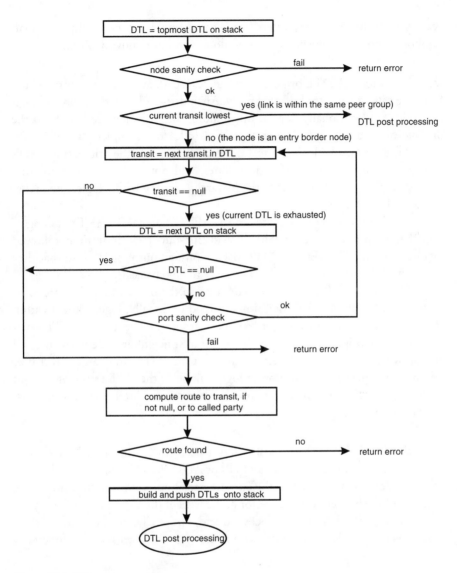

Figure 7.26 DTL preprocessing.

tor nor entry border node, (4) DTL terminator, (5) exit border node, and (6) neither DTL terminator nor exit border node.

A call exits from a peer group through an exit border node and enters to a peer group at an entry border node. A call is said to be *entering* a peer group at a level of the hierarchy if it has been progressed to a border node in that peer group, whereas it is said to be *exiting* a peer group at a level of the hierarchy if a call is to be progressed next by a border node in another peer group. The processing at the border nodes differs from that at nonborder nodes. In particular, nonborder nodes, unlike border nodes, do not add or remove a DTL.

DTL Originator. A DTL originator calculates a path across the routing domain from the calling party to the called party. An end-to-end path comprises all logical nodes (and optionally, logical links) along the path that are known to the originating node. The end-to-end path includes the complete list of logical nodes (and optionally, logical links) necessary to traverse the originating node's own peer group and a list of logical nodes (and optionally logical links) that specifies a path at higher levels of the hierarchy, ending with a logical node that contains the DTL terminator.

The DTL originator appends to the SETUP or ADD PARTY message a broadband repeat indicator IE with "last-in-first-out-stack" indication, followed by one or more DTL IEs (i.e., DTL IEs are pushed onto the stack so that they appear in the reverse order in which they are to be traversed).

The top DTL IE in the stack contains, in the order which they are to be traversed, a list of all of the logical nodes (and optionally, logical links) within the DTL originator's own lowest level peer group. Hence, the top DTL starts with the DTL originator, followed by the nearest neighbor node in the list and ending with either the DTL terminator (if in the DTL originator's own peer group) or the peer group exit border logical node (if the DTL terminator is not in the DTL originator's own peer group). The transit pointer is set to 1, indicating the originating node.

Entry Border Node. When a call enters a peer group, the entry border node examines the top DTL IE in the SETUP or ADD PARTY message. If the logical node identifier indicated by the current transit pointer is not the same as the node ID of the border node's ancestor at the level of the common peer group for the receiving link, then the call is cleared and, optionally, cranked back. Otherwise, the entry border node calculates a route across the peer group as follows:

1. It examines the DTL at the top of the stack. If the current transit pointer indicates the end of the DTL, it performs the procedures in (2). Otherwise, it proceeds with the procedures in (3).

2. Upon reaching the end of a DTL, the node examines the next lower DTL on the stack, if there is one. If the logical node identifier indicated by the current transit pointer of the next DTL is not a node ID of one of the border node's ancestors, the call is cleared and, optionally, cranked back:

- If the current transit pointer indicates the end of that DTL and there is a DTL lower on the stack, the procedures of (2) are recursively repeated.
- If the current transit pointer does not indicate the end of the DTL, it proceeds with the procedures in (3), using the transit indicated by the current transit pointer as the target of the path calculation.
- If there is not a DTL lower on the stack, the entry border node saves the level of this DTL (at the bottom of the DTL stack) as the path scope before determining whether the node is the DTL terminator.
- Otherwise, it proceeds with the procedures in (3), using the called party number as the target of the path calculation. If a *transit network selection* IE is present, it includes the designated transit network along the route.

3. The entry border node calculates the path at the lowest level in which the entry border node appears. DTLs describing the calculated path are pushed onto the DTL stack in the reverse order in which they are to be traversed. Each DTL contains a list of transits at a single level of the hierarchy. The current transit pointers in these DTLs are set to 1, indicating the first transit. If no suitable path exists to the target of the path calculation, then the call is cleared or cranked back.

Neither DTL Originator nor Entry Border Node.

If a node is not a DTL originator or an entry border node, the top DTL IE in the SETUP or ADD PARTY message is examined. If the node identifier indicated by the current transit pointer is not the same as that node's own node identifier, the call is cleared and, optionally, cranked back. After processing the signaling message and performing call admission control, if the node accepts the connection request, the SETUP or ADD PARTY message is transmitted to the next node along the path, as specified in the DTL.

If the current transit pointer in the top DTL IE on the DTL stack does not point to the last transit in the DTL IE, the current transit pointer is advanced. If the current transit pointer indicates a transit that is a neighbor at the same level of hierarchy, the connection is progressed to that transit, using the specified logical port (if any). If the specified logical port does not correspond to a logical link to the next transit or the current transit pointer indicates a transit that is not a neighbor, the connection is cranked back.

DTL Terminator.

If the current transit pointer indicates the end of the DTL, the DTL is popped from the stack and the level of the DTL is saved as the path scope. If there are any DTLs remaining on the stack, the node examines the next DTL. If the logical node identifier indicated by the current transit pointer of the next DTL is not a node ID of one of the border node's ancestors, then the call is cleared and, optionally, cranked back.

If the DTL stack becomes empty, this node is the DTL terminator. If a transit network selection IE is present and the indicated transit network is served by this node, then the node progresses the call to the transit network using UNI procedures. If there is no transit network selection IE and the called party is served by this node, the node progresses the call to the called party using the UNI procedures. Otherwise, if the DTL stack becomes empty but the called party or transit network is unreachable, then the call is cleared or cranked back.

Neither DTL Terminator nor Exit Border Node.   The procedures at nodes that are neither DTL terminators nor exit border nodes are the same as the procedures at nodes that are neither DTL originators nor entry border nodes.

Exit Border Node.   If the next node along the path is an entry border node, the processing at an exit border node is the same as it is in a neither DTL originator nor an entry border node. An exit border node may also provide connectivity to an exterior route. An exterior route is by definition a route that leaves the PNNI routing domain. It is not feasible to specify DTLs for the exterior routes. For example, in most cases, the public networks that are managing the exterior networks will not want to expose any internal topology within their network. Furthermore, there is no guarantee that all PNNI routing domains worldwide will use a consistent definition of hierarchical levels. In this case, if a DTL was used for an exterior route, it might not be possible for the DTL to make use of consistent level indicators along the path. Therefore, if a path makes use of an exterior route, the DTL terminates at the node that advertises the exterior route into the PNNI routing domain.

7.5.4.3 Summary of DTL Processing

The operations required to process the received stack of DTLs and to generate the new stack of DTLs are summarized in Figure 7.26 (shown earlier) and Figure 7.27.

Upon receiving a PNNI SETUP message, the node extracts the topmost DTL stack from the SETUP or ADD PARTY message and checks whether or not the current transit in the topmost DTL on the stack corresponds to itself at any level of the routing hierarchy. If it does not, it is an error condition.

If the transit corresponds to the current node at the lowest level of the hierarchy, the target node could be a peer node to which a direct connectivity exists or to the destination end station to which the current node knows the UNI the station is attached to. If the target node is a logical node (at a higher

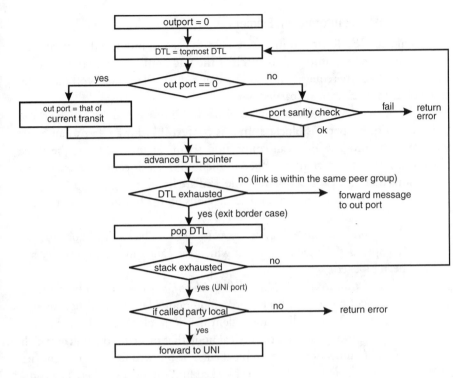

Figure 7.27 DTL postprocessing procedures.

level of the routing hierarchy than the lowest level), the current node needs to determine a route from itself to the target node. Once the route is determined based on the current node's view of the network, a new DTL specifying the route is pushed onto the stack. After processing the current transit, the current node pops completed DTLs from the stack and pushes the new DTLs. It then advances the current transit pointer to the next DTL in the stack.

7.5.5 Crankback

Crankback is the process of attempting to reroute a call to an alternate path within a PNNI domain. The switching system that rejects the call initiates a call-clearing message, which traverses the end-to-end path between the originating switch and the switch that rejected the call in the reverse order. As the call-clearing message travels towards the originating switch, one or more intermediate switches along the path may attempt to reroute the call using an alternate path. Doing so, if it results in a successful call setup, might reduce the end-to-end connection setup time. However, not every rejected call request is cranked back (i.e., any call rejected by the called user is not cranked back).

7.5.5.1 Crankback Information Element

The crankback IE, illustrated in Figure 7.28, is used to indicate that crankback procedures have been initiated. It specifies the node or link where the call/connection or add party request is not accepted and the level of the PNNI hierarchy at which crankback is being carried out.

Crankback level indicates the level of the PNNI hierarchy at which the call/connection or party is blocked. In this context, blocking refers to an event that stops the progress of a call/connection setup. Figure 7.29 lists the three blocked transit types currently defined in the PNNI specification, whereas various reasons that may cause a call to be blocked are listed in Figure 7.30.

The blocked transit identifier field depends on the blocked transit type:

- *Blocked transit type = blocked node identifier*—The blocked node identifier specifies the logical node at which the call/connection or party has been blocked.
- *Blocked transit type = blocked link identifier*—Blocked link identifier consists of blocked link's (1) preceding node identifier, (2) port identifier, and (3) succeeding node identifier. The blocked link's preceding node identifier specifies the logical node that precedes the link at which the call/connection has been blocked. The blocked link's port identifier specifies the logical port of the blocked link's node identifier. The combination of the blocked link's node identifier and the blocked link's port identifier unambiguously identifies the link at which the call/connection has been blocked. If a port identifier is set to 0, all links should be considered blocked. Finally, the blocked link's succeeding node

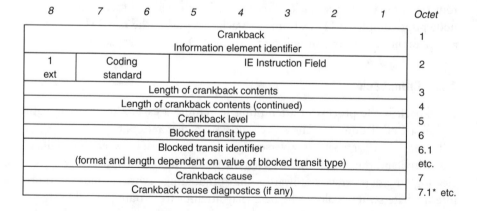

8	7	6	5	4	3	2	1	Octet
Crankback Information element identifier								1
1 ext	Coding standard	IE Instruction Field						2
Length of crankback contents								3
Length of crankback contents (continued)								4
Crankback level								5
Blocked transit type								6
Blocked transit identifier (format and length dependent on value of blocked transit type)								6.1 etc.
Crankback cause								7
Crankback cause diagnostics (if any)								7.1* etc.

Figure 7.28 Crankback IE.

8 7 6 5 4 3 2 1	Meaning
0 0 0 0 0 0 1 0	Call or party has been blocked at the succeeding end of this interface
0 0 0 0 0 0 1 1	Blocked node indicator
0 0 0 0 0 1 0 0	Blocked link indicator

Figure 7.29 Blocked transit type field values.

Number	Meaning
2	Transit network unreachable
3	Destination unreachable
32	Too many pending add party requests
35	Requested VPI/VCI not available
37	User cell rate not available
38	Network out of order
41	Temporary failure
45	No VPI/VCI available
47	Resource unavailable, unspecified
49	Quality of Service unavailable
57	Bearer capability not authorized
58	Bearer capability not presently available
63	Service or option not available, unspecified
65	Bearer service not implemented
73	Unsupported combination of traffic parameters
128	Next node unreachable
160	DTL Transit not my node ID

Figure 7.30 Crankback causes.

identifier identifies the logical node at which the call/connection or party has been blocked. The crankback cause field specifies the reason for the crankback, as illustrated in figure 7.30.

The use of crankback cause diagnostics and the coding format of the diagnostics vary among the different cause values. Currently, only causes 37 and 49 have diagnoses defined. For cause 37, the cause diagnostics field may optionally include updated topology state parameters for PNNI generic call admission control (GCAC). For cause 49, four logical fields are defined for cause code: cell transfer delay, cell delay variation, cell loss ratio, and other QoS. If the corresponding bit is set to 1, then the requested QoS for that parameter is available, while it is not available if the bit is set to 0

7.5.5.2 Crankback Request

In general, there are four cases that may cause a crankback: (1) reachability errors, (2) resource errors, (3) DTL processing errors, and (4) policy violations. Policy violations are not currently specified, and are left to be addressed in the future releases of the PNNI specification. The other three cases are discussed in the following subsections.

Reachability Errors

This type of failure indicates that no path within the scope of the received DTL stack exists to the destination (which is different than rejecting a call request due to lack of resource availability). When crankback occurs due to reachability errors, the blocked transit specified in the crankback IE must be a blocked link.

If the next node, transit network, or called party address cannot be reached from this node through any logical link, the blocked link's port ID is set to 0. If the unreachability occurs at a port specified in the DTL stack, the blocked link's port ID shall be set to that port ID.

In the case of unreachable transit networks and called party addresses, the blocked link's succeeding node ID is set to 0. There are two types of cause codes defined for reachability errors.

Destination Unreachable and Transit Network Unreachable. This cause code is returned when a node receives a SETUP or an ADD PARTY message with a DTL stack indicating that it is the last node in the path and the called party or transit network is not reachable from that node.

It is possible that the node's link state database includes reachability information for the called party or the transit network through different nodes. This may be the case when a call enters a partitioned peer group, for example, due to link/node failures. The crankback in this case may be initiated as the destination may possibly be reached via some other partition.

In the case of partitioned peer groups, it is possible that more than one logical node advertises reachability to a given destination. For example, a route may consist of logical nodes A and B, in that order. If B is partitioned so that the destination is not reachable from the partition, the call-establishment request enters network A. It may be possible to reroute the call so that the entry node at B is in the partition that has connectivity to the destination.

When this crankback cause code is specified, the node that generated the DTL may or may not be able to see the other partition(s). Alternate routing can be attempted for the call if the other partition(s) are seen, whereas the call can be cranked back further if they are not seen by the node that generated the DTL.

Next Node Unreachable.   This cause code is returned when the next transit in the DTL stack is not directly reachable from this node. The cause code is set to "no route to specified transit network," if a transit network selection IE is present, or "no route to destination" otherwise.

Resource Errors

The resources needed to support a call are calculated from the traffic and QoS parameters included in the SETUP or ADD PARTY messages. Resource errors are used to signal that a path could not be found for the call to satisfy the requested QoS parameters with the specified traffic parameters.

Inability to satisfy the requested service can be detected either during GCAC/path calculation or during actual CAC. Calls that are rejected in a PNNI domain due to insufficient resources are always cranked back.

If the requested user cell rate(s) from the ATM traffic descriptor IE cannot be satisfied, the call is cranked back with cause "user cell rate not available." If no path can be found to satisfy the requested maximum CTD, peak-to-peak CDV, and/or CLR (in one and/or the other direction for a call), the call is cranked back with cause "QoS unavailable." The specific QoS parameter(s) that caused the call rejection are indicated in the diagnostics by setting the appropriate bits to "CTD unavailable," "CDV unavailable," and/or "CLR unavailable." Connection identifiers are also resources used by connections, and a resource error due to unavailability of VPIs/VCIs may also cause a crankback.

If blocking happens due to insufficient resources at the succeeding end of the previous link, calls requesting similar resource requirements might be accepted on other ports. Similarly, if call rejection is due to insufficient resources at the preceding end of the following link, calls requesting similar resource requirements on other ports might be accepted. However, if resources are not available within the node itself, then all calls requesting similar resource requirement from this node are likely to be blocked. The PNNI signaling procedures are different for each of these three cases, as discussed in subsequent sections.

DTL Processing Errors

When a node receives a DTL IE in which the DTL identified by the current transit pointer does not correspond to the node at any level of the hierarchy, crankback may occur. In this case, the crankback IE includes either a succeeding end blocked indicator with cause "next node unreachable" or the DTL transit is listed as the blocked node with cause "DTL transit not my node identifier."

7.5.5.3 Procedures for Crankback Level and Blocked Transit

The crankback procedures provide the mechanisms required to manage a crankback from the point of blocking, through intermediate nodes, to one or more nodes that are allowed to choose alternate routes for the call. The crankback level and blocked transit type are used during crankback to determine which nodes are allowed to choose alternate routes or alternate links, respectively. The blocked node or link must be avoided in all alternate routes attempted for this call.

Any node that might take part in alternate routing must keep the copies of the originally received SETUP message contents (until a CONNECT/ALERT-ING message is received or party is cleared). The saved contents include the QoS and traffic parameters used in routing, call admission control decisions, and any DTLs generated at that node. Copies of any blocked node or blocked link subfield contents received by the node during crankback procedures are also kept.

Procedures at the Point of Blocking

The procedures carried out at the point of blocking vary, depending on whether crankback occurs due to problems at the input port, at the output port, or within the node itself.

Blocking at a Node

Upon blocking at a node, crankback procedures are initiated by sending an appropriate call/connection-clearing message with a crankback IE. The crankback IE must include a crankback level subfield whose value is set to the level of the first node identifier indicated in the top DTL. The blocked transit type is set to "blocked node" and the blocked transit identifier indicates the node's own node identifier at the corresponding level of hierarchy, indicated by the current transit pointer in the top DTL on the stack in the received SETUP or ADD PARTY message. The crankback IE also contains the crankback cause subfield.

Blocking at the Preceding End of a Link

Link blocking can be determined at the preceding end of a link when CAC within the node determines that sufficient resources are not available. If the node at the preceding end of the link is not an entry border node for the call, and if there are other links that satisfy the DTLs in the signaling message, then alternate routing may be attempted by this node.

If no alternate routing is attempted or if an alternate routing attempt fails, then the node at the preceding end of the link must crank back the call by sending an appropriate clearing message with a crankback IE. This information element includes a crankback level subfield whose value is set to the level of the first node ID indicated in the top DTL of the stack. The blocked transit type subfield is set to "blocked link" and the blocked transit identifier indicates the identity of the blocked link. The blocked link's preceding node identifier and port identifier are set to the node and port identifiers indicated by the current transit pointer in the top DTL on the stack. The blocked link's succeeding node identifier is set to the next node identifier in the top DTL on the stack.

Blocking at the Succeeding End of a Link

Link blocking can be determined at the succeeding end of a link when CAC within the node determines that sufficient resources are not available. In this case, the node must crank back the call by sending an appropriate clearing message with a crankback IE. The crankback IE must include a crankback level subfield whose value is set to the level of the first node ID indicated in the top DTL of the stack. The blocked transit type must be set to "call or party has been blocked at the succeeding end of this interface." The crankback IE also contains the crankback cause subfield.

Receiving a Clearing Message With a Crankback Information Element

Upon receiving a clearing message (RELEASE, RELEASE COMPLETE, or ADD PARTY REJECT) including a crankback IE, a node first checks whether the blocked transit type indicates that the "call or party has been blocked at the succeeding end of this interface." If this is the case, the procedures for receiving a clearing message indicating blocking at this interface are followed. If this is not the case, and if the node generated any DTLs for this call of equal or higher level than the crankback level, then the procedures for receiving a clearing message at the entry border node at the crankback level are followed. Otherwise, the node must crank back the call or party by sending an appropriate clearing message (RELEASE or ADD PARTY REJECT) including an unchanged crankback IE over its previous interface (towards the calling party).

Receiving a Clearing Message Indicating Blocking at This Interface

If the node at the preceding end of the link is an entry border node for the call, the entry node procedures discussed next apply. Otherwise, if other links exist that satisfy the DTLs in the SETUP or ADD PARTY message received by this node, then alternate routing may be attempted. This may happen when there are multiple links between the two nodes. In addition, a SETUP message may only be resent on the blocked link with a different VPI (for SVPs) or VPI/VCI pair (for SVCs) if crankback cause "requested VPI/VCI not available" is present.

If no alternate routing is attempted or if alternate routing fails, then the node must continue to crank back the call or party. The blocked transit type subfield must be changed to indicate "blocked link," and a blocked transit identifier must be inserted into the crankback IE.

Receiving a Clearing Message at the Entry Border Node at the Crankback Level

The node that generated the DTL at the crankback level has been reached. This node must determine whether to try alternate routing or to continue cranking back the call or party. If alternate routing is attempted, then the call shall be progressed according to the procedures of receiving a clearing message indicating blocking at this interface.

Alternate Routing

If alternate routing is attempted, the routing computation must produce a path consistent with the DTLs in the originally received SETUP or ADD PARTY message that does not contain any blocked nodes and/or links received in the crankback IE of any clearing messages (RELEASE, or ADD PARTY REJECT). Except as a result of those procedures normally undertaken by entry border nodes, the SETUP or ADD PARTY message progressed should be similar to the originally received SETUP or ADD PARTY message.

Crankback to Next Higher Level

If alternate routing is not attempted or if alternate routing fails to find a suitable path across the peer group, then crankback must proceed to the entry border node of the next higher level peer group. The crankback level must be set to the level of the first node in the top DTL on the stack in the received SETUP or ADD PARTY message. The identity of the blocked node or link in the crankback IE must be changed to reflect a node or link known to the parent peer group, as opposed to a node or link known to the child peer group.

If only nodes and/or links internal to the peer group were returned as blocked nodes or links in the crankback IE of RELEASE or ADD PARTY REJECT messages, then the logical node corresponding to the peer group should be listed as blocked. In this case, the blocked node ID should match the node ID indicated by the current transit pointer in the top DTL on the stack in the received SETUP or ADD PARTY message.

7.5.5.4 Carrying Updated Topology State Parameters in the Crankback Information Element

Blocking due to resource errors often occurs due to the use of out of date or inaccurate values of routing parameters. To reduce the impact of this uncertainty, the GCAC routing parameters can be optionally included as diagnostics when the crankback cause is "user cell rate not available." The updated GCAC parameters apply to the same service category as in the connection request. The updated parameter values are included for one direction only, since each node is responsible for managing the values of the routing parameters it owns (i.e., the outgoing direction on its links).

A local node always has the best knowledge of its topology state parameters and the last advertised values may or may not match the current values. When forming the crankback IE with a resource error, the node initiating crankback of the call may include the most current values of various resource parameters.

In addition to the updated parameter values, a direction and port ID are specified. For blocked nodes, the direction subfield is coded as "forward" and the port ID is coded as "zero." For blocked links, the forward direction indicates that the call was blocked at the preceding side of a link, whereas the backward direction indicates that the call was blocked at the succeeding side of the link. If the direction is set to "forward," the updated topology state parameters apply to the outgoing direction of the specified port, from the blocked link's preceding node. If the direction is set to "backward," the port ID is set to 0 by the node initiating crankback of the call. The updated parameter values are those values that apply to the physical link or VPC over which the SETUP or ADD PARTY message was received. When the preceding node receives the crankback IE with the "succeeding end blocked" indicator, it must insert a port ID that identifies its port for the PNNI interface. For uplinks, the updated parameter values apply to the incoming direction of the specified port on the blocked link's preceding node. For horizontal links, the updated parameter values apply to the outgoing direction on the corresponding port on the blocked link's succeeding node.

7.5.5.5 Summary of Crankback Procedures

Figure 7.31 illustrates the crankback procedures.

7.5.6 Restart Procedure

The restart procedure is used to return a switched virtual channel, all virtual channels in a virtual path, a switched virtual path, or all switched virtual channels and switched virtual paths controlled by the signaling virtual channel to the idle condition. The procedure is usually invoked when the other side of the interface does not respond to call control messages or a failure has occurred (e.g., following the expiry of timer T308 due to the absence of response to a clearing message). It may also be initiated as a result of local failure, maintenance action, or misoperation.

PNNI restart procedures are the same as in Q.2931, with minor changes. In particular, both sides of the PNNI interface are required to implement the restart procedures and a RESTART message may be sent by either side of PNNI in order to return virtual channels to the idle condition.

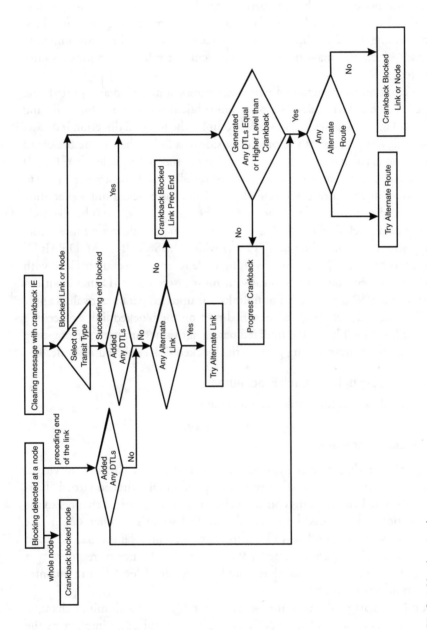

Figure 7.31 Crankback procedures.

A restart collision occurs at PNNI when signaling entities on both sides of the interface simultaneously transmit a RESTART message. The call reference flag of the global call reference applies to restart procedures. In the case when both sides of the interface initiate simultaneous restart requests, they shall be handled independently. When the same virtual channel(s) are specified, they can not be considered free for reuse until all the relevant restart procedures are completed.

Figure 7.32 illustrates the message exchange for the PNNI restart procedure. Upon sending a RESTART message, the side that initiated the procedure starts timer T316 and starts waiting for the RESTART ACKNOWLEDGE message. If T316 expires before the reply is received, or upon receiving the acknowledgment, the initiating side clears the call internally. The side that received the RESTART message starts initiating the internal call-clearing procedures and sends a RESTART ACKNOWLEDGE message back to indicate that the call is cleared.

7.5.7 Traffic Parameter Selection Procedures for ABR Connections

PNNI signaling uses the UNI 4.0 network side ABR procedures to set up an ABR connection at the PNNI. The use of ABR setup parameters are mandatory in the PNNI SETUP message. The originating switching system is responsible for sending the SETUP message with all of the required parameters in the case of soft PVPC or PVCC setup. Default parameter values used for ABR connections at a PNNI are the same as the ones defined in UNI 4.0.

The procedures to support ABR service are described in Chapter 4 and are not repeated here. A summary of various features of signaling procedures for the ABR is given instead. ABR parameter values for a given direction can be negotiated by either side of the PNNI. MCR is negotiated if the corresponding MCR parameter is included in the minimum acceptable ATM traffic descriptor IE in the SETUP message. If able to provide the indicated PCR and ABR setup pa-

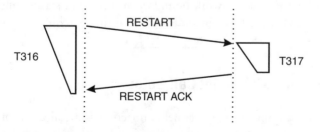

Figure 7.32 PNNI restart message exchange.

rameter values, the PNNI network progresses the call towards the called user, with the original parameters. When progressing the call, the network may, if necessary, also adjust either or both forward and backward ABR setup parameters: ABR initial cell rate (ICR), ABR transient buffer exposure (TBE), rate increase factor (RIF), and rate decrease factor (RDF).

A PNNI node is required to maintain the following relation between MCR, ICR, and PCR:

$$MCR \leq ICR \leq PCR$$

If a PNNI node cannot provide the indicated PCR, but it is able to provide at least the MCR value as negotiated, the call progresses towards the called user after adjusting the PCR value. The adjusted PCR value is required to be greater than or equal to the MCR value. If the network is not able to support a peak cell rate greater than or equal to MCR, the crankback procedures are followed with cause and crankback cause "user cell rate not available."

7.5.7.1 Processing of Cumulative RM Fixed Round-Trip Time Parameter for ABR Connections

At each switching system, the cumulative ABR RM fixed round-trip time parameter in the *ABR setup parameters* IE is adjusted by the round-trip time contribution of the previous hop (UNI or PNNI). The final switching system is an exception: it includes both the previous hop and the next hop (to the called user) in its adjustment. The parameter is adjusted by the succeeding side. The amount of the adjustment is the sum of the forward and reverse link propagation delays and any fixed processing delays at the PNNI (including the switching delays).

In the first and last switching systems of the call path, the network side of the UNI adjusts the cumulative RM fixed round-trip time parameter. The amount of the adjustment is the sum of the forward and reverse link propagation delays between the network boundary and the end station and any fixed processing delays at the UNI, including those at the switching system.

7.6 A Point-to-Multipoint Call Setup

A point-to-multipoint virtual channel connection is a collection of associated virtual channel links that connect one end point (root) to two or more end points (leaves). This capability only supports unidirectional transport from the root to the leaves. Leaves can be added and removed during the lifetime of a connection.

7.6.1 First Party Setup

Upon reception of an add party request at the PNNI network interface for the first party of a point-to-multipoint call, the connection setup proceeds by transferring a SETUP message (as opposed to ADD PARTY) across the PNNI interface. The SETUP message shall contain the endpoint reference IE and a broadband bearer capability IE indicating point-to-multipoint in the user plane connection configuration field (see Figure 7.33). The network includes AAL parameters and the broadband low-layer IEs in the SETUP message when the originating user included these information elements in the SETUP or ADD PARTY message for this party.

If the CALL PROCEEDING message sent in response to a SETUP message does not contain an endpoint reference IE or there is a content error (e.g., incorrect endpoint reference value or flag), it is treated as a mandatory information element content error.

Upon transmitting the SETUP message, the preceding side enters the add party initiated party state. When receiving a SETUP message, with the user plane connection configuration field in the broadband bearer capability IE set to point-to-multipoint, the succeeding side enters the add party received party state.

Upon transmitting the ALERTING message, the succeeding side enters the party alerting delivered party state. Upon reception of the ALERTING message, the preceding side enters the party alerting received party state.

Upon sending or reception of a CONNECT message, the PNNI enters the active party state.

Party state	Call/connection party	*PRECEDING SIDE*	*SUCCEEDING SIDE*	Call/connection party	Party state
P1	NN6	SETUP →		NN1 NN3	P2
	NN9	← CALL PROCEEDING		NN4	P3
P4	NN7	← ALERTING		NN10	P7
P7	NN10	← CONNECT			

Figure 7.33 Point-to-multipoint setup.

Upon sending or reception of a CALL PROCEEDING message, the PNNI does not change party state.

7.6.2 Adding a Party

Upon reception of an add party request, the preceding side transfers an ADD PARTY message across the PNNI, starts timer T399, and enters the add party initiated party state (see Figure 7.34). The PNNI network transfers the ADD PARTY message only if the link is in the active or alerting received link state.

Upon receipt of an ADD PARTY message, the succeeding side enters the add party received party state. If there is one and only one party in the add party initiated party state and the call is not in the active or alerting received link state, additional add party requests are queued by the preceding side on the add party queue until the link state becomes alerting received, active, null, release indication, or release request. At this point, the queued add party requests are treated as if they had just arrived.

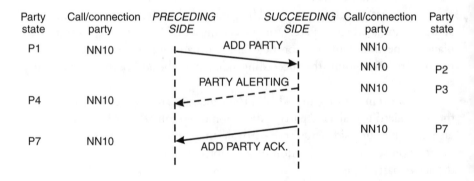

Figure 7.34 Adding a party to an already established point-to-multipoint connection.

7.6.3 Messages for ATM Point-to-Multipoint Call and Connection Control

7.6.3.1 ADD PARTY

The ADD PARTY message is used to extend a previously established point-to-multipoint connection to a new called party. The processing of this message is the same as the PNNI SETUP message. However, if the next node specified in the DTL is already in the tree (i.e., the point-to-multipoint connection already includes that branch of the tree), there is no need for the node to reserve/allocate any additional resources. For each called party, a PNNI node maintains a sub-call state for call management (i.e., establishment, activation, and termination).

An endpoint reference IE is used to identify the end point for this call locally at an interface. In particular, this information element is used together with the call reference to associate the called party to the call at a particular interface. The endpoint reference must be unique within a given call reference on a given link.

If the point-to-multipoint connection doesn't exist between the current node and the next node along the path as defined in the DTL IE, the ADD PARTY message is translated to a SETUP message and transported to the next node. This SETUP message is different in that the traffic parameters are not negotiable and the QoS parameters are the same as the ones agreed upon during the call establishment of the first leaf. The only exemption is the CTD (i.e., carried in the end-to-end transit delay), which may differ for each leaf.

Calling party soft PVPC/PVCCs, called party soft PVPC/PVCCs, broad-band repeat indicators, and DTL IEs have the same use as defined in the SETUP message (see Figure 7.35).

7.6.3.2 ADD PARTY ACKNOWLEDGE

An ADD PARTY ACKNOWLEDGE message is used to indicate the ADD PARTY request is accepted. ADD PARTY ACKNOWLEDGE is similar to the CONNECT message used in point-to-point call setup except that it does not include any negotiated parameters, as the ADD PARTY message does not have any means for negotiation. The end-to-end transit delay IE in this message includes the actual CTD from the root to the leaf, which is less than or equal to the value sent in the ADD PARTY message (see Figure 7.36).

The endpoint reference must be the same value as in the ADD PARTY message to which it is responding.

7.6.3.3 PARTY ALERTING

PARTY ALERTING procedures for point-to-multipoint are the same as they are for point-to-point connections. The endpoint reference is included to uniquely identify the leaf locally across a PNNI (see Figure 7.37.)

Information Element	Type	Length
AAL parameters	O[1]	4–21
Broadband high-layer information	O[1]	4–13
Broadband low-layer information	O[1]	4–17
Called party number	M	£25
Calling party soft PVPC/PVCC	O	4–10
Called party soft PVPC/PVCC	O	4–11
Called party subaddress	O[1]	4–25
Calling party number	O[1]	4–26
Calling party subaddress	O[1]	4–25
Broadband repeat indicator	M	5
Designated transit list	M	33–546
Endpoint reference	M	7
End-to-end transit delay	O[1]	4–12
Notification indicator	O[1]	4–*
Transit network selection	O[1]	4–8

[1] Included if the received add party indication contains this information.

Figure 7.35 Information elements used in the ADD PARTY message.

Information Element	Type	Length
Broadband low-layer information	O	4–17
AAL parameters	O	4–20
Called party soft PVPC/PVCC	O	4–11
Connected number	O	4–26
Connected subaddress	O	4–25
Endpoint reference	M	7
End-to-end transit delay	O	4–7
Notification indicator	O	4–*

Figure 7.36 Information elements used in the ADD PARTY ACKNOWLEDGE message.

A notification indicator is included if the received party alerting indication contains this information.

7.6.3.4 ADD PARTY REJECT

This message is used to indicate that the ADD PARTY request is not accepted (see Figure 7.38). This message corresponds to the RELEASE COMPLETE

Information Element	Type	Length
Endpoint reference	M	7
Notification indicator	O	4–*

Figure 7.37 Information elements used in the PARTY ALERTING message.

Information Element	Type	Length
Cause	M	6–34
Crankback	O	4–72
Endpoint reference	M	7

Figure 7.38 Information elements used in the ADD PARTY REJECT message.

message when it is used to reject a call setup. The point-to-multipoint procedures also include the crankback capability, which is performed the same way as point-to-point calls.

A *crankback* IE may be included to indicate crankback. The endpoint reference must be the same value as in the ADD PARTY message to which it is responding.

7.6.3.5 DROP PARTY

The DROP PARTY message is sent to clear a party from an existing point-to-multipoint connection (see Figure 7.39). It is processed similar to the RELEASE message processing. The DROP PARTY message is converted to a RELEASE message if the dropped party is the last one of the point-to-multipoint connection established at that interface.

A *notification indicator* IE is included if the received drop party indication contains this information.

7.6.3.6 DROP PARTY ACKNOWLEDGE

This message acknowledges that the resources associated with the endpoint reference (which uniquely identifies a leaf) are released. This message itself does

Information Element	Type	Length
Cause	M	6–34
Endpoint reference	M	7
Notification indicator	O	4–*

Figure 7.39 Information elements used in the DROP PARTY message.

not cause any resources associated with the call to be released. If there is no active party associated with this call across that PNNI interface, the call will be cleared at that interface using the RELEASE message. Otherwise, if there is at least one party associated with this call across that PNNI interface, the DROP PARTY ACKNOWLEDGE causes only the endpoint reference to be released and the associated substate machine to become null (see Figure 7.40).

A cause IE is mandatory when DROP PARTY ACKNOWLEDGE is sent as a result of an error condition.

7.6.4 Crankback Procedures for Point-to-Multipoint Calls/Connections

The procedures used to process clearing messages with crankback for point-to-multipoint calls/connections are the same as those for point-to-point calls/connections, except when a RELEASE or RELEASE COMPLETE message is received and there are queued ADD PARTY requests on the ADD PARTY queue.

If the network node does not attempt alternate routing for the party for which the RELEASE or RELEASE COMPLETE message was received, and the blocked transit in the crankback IE is the succeeding end of this interface, then the network node sends an ADD PARTY REJECT message for each queued ADD PARTY request towards the preceding network node, with the crankback indication according to the previous subsections.

If the network node does not attempt alternate routing for the party for which the RELEASE or RELEASE COMPLETE message was received and the blocked transit in the crankback IE is not the succeeding end of this interface, then the network node cranks back the party for which the RELEASE or RELEASE COMPLETE message was received, and does one of the following:

1. Progresses one of the ADD PARTY requests on the ADD PARTY queue by sending a SETUP message, leaving the remaining ADD PARTY requests pending;
2. Cranks back all ADD PARTY requests on the ADD PARTY queue whose DTLs contain the blocked transit, and progresses one of the ADD PARTY requests remaining on the ADD PARTY queue (if any)

Information Element	Type	Length
Cause	O	4–34
Endpoint reference	M	7

Figure 7.40 Information elements used in the DROP PARTY ACKNOWLEDGE message.

by sending a SETUP message, leaving the remaining ADD PARTY requests pending.

If alternate routing is attempted for the party for which the RELEASE or RELEASE COMPLETE message was received, the node shall determine if and how the queued ADD PARTY requests can be satisfied. If the request can be satisfied by adding it to a branch that is in the active state, it sends an ADD PARTY message to its succeeding node. If a new branch is required and the branch is in the null state, it sends a SETUP message to the corresponding succeeding node. If a new branch is required and the branch is in the call initiated, outgoing call proceeding, or call delivered state, it places the party in the processing queue. More than one branch may be needed to satisfy all the queued ADD PARTY requests. If an ADD PARTY request cannot be satisfied, the network node sends an ADD PARTY REJECT message for that ADD PARTY request toward the preceding network node, with the crankback indication according to the previous subsections. The manner in which the node determines whether an ADD PARTY request can be satisfied, and how it will do so, is implementation specific.

7.7 Soft Permanent Virtual Connections

Consider two end stations connected to a PNNI network via either a permanent virtual channel connection (PVCC) or a permanent virtual path connection (PVPC). These connections at the UNI are established offline (i.e., via the management interfaces). When these two end stations want to communicate with each other, an end-to-end path needs to be established across the PNNI network. This capability in PNNI is referred to as soft PVPC/PVCC, as the end-to-end connection consists of three connections glued to each other: two connections, one each at the interface between an end station and the network, and one switched connection in the PNNI network providing connectivity between the two UNIs. In the case of point-to-multipoint connections in which all end stations to the network are connected via permanent connections, the soft permanent virtual connections capability allows the PNNI network to provide a point-to-multipoint switched connection, providing connectivity from the root to all the leaves.

A soft PVPC/PVCC is established and released between the two network interfaces (NIs) serving the permanent virtual connection. The NIs are the end points of a soft PVPC/ PVCC, and they are identified by unique ATM addresses (including the selector byte). One of the two end points of a PVPC/PVCC owns the connection and it is responsible for its management.

This network interface is referred to as the calling end point. It is also the responsibility of the calling end point to try to reestablish the connection if the switched portion of the PVPC/PVCC gets disconnected due to switching system or link failures. Figure 7.41 illustrates the concept of soft connections.

The address of the calling end point is encoded in the calling party number IE, whereas the address of the called party (i.e., destination end point) is encoded in the called party IE. The ATM addresses of the end points of a soft PVPC/PVCC as well as the VPI/VCIs used between the NIs and the end stations are provided by the network management system.

The procedures used to establish point-to-point and point-to-multipoint soft PVPC/PVCCs are essentially the same as point-to-point and point-to-multipoint procedures for switched virtual connections. The only difference is that various information elements in the PNNI signaling messages are used to carry end-to-end information for end stations to process for switched connections at the UNI. As the end stations connected to the network in this case don't support signaling (i.e., use permanent connections to attach to the network), these information elements don't need to be carried across the PNNI network for soft PVPC/ PVCCs. In particular, AAL parameters, broadband high-layer information, broadband low-layer information, called party subaddress, and calling

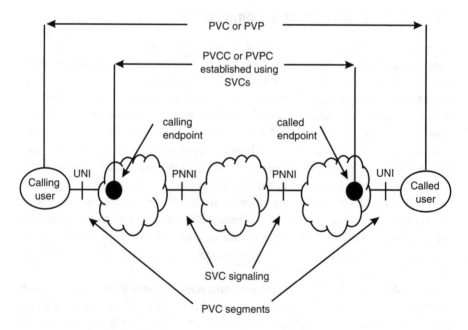

Figure 7.41 The concept of soft connections.

party subaddress information are not included in the SETUP, CONNECT, ADD PARTY, and ADD PARTY ACKNOWLEDGE messages.

The parameters of the PVPC or PVCC are established administratively. Accordingly, the negotiation of end-to-end traffic and QoS parameters is not possible with soft PVPC/PVCCs. If any of these parameters specified in the SETUP message cannot be supported by the switching system, the call is cranked back with cause "resources not available, unspecified."

7.7.1 PVPC/PVCC Information Elements

As discussed previously, the address of the calling end point is encoded in the calling party number IE, whereas the address of the called party (i.e., destination end point) is encoded in the called party IE.

7.7.1.1 Calling Party Soft PVPC or PVCC Information Element

The calling party soft PVPC or PVCC IE (Figure 7.42) is used to indicate the VPI (in the case of PVPC) or VPI/VCI (in the case of PVCC) values used for the PVC segment between the calling connecting point and the user of a PVPC or PVCC. These values are conveyed to the called connecting point transparently by the intermediate PNNI nodes along the connection's path.

The VPI value field carries the VPI value assigned to the connection. Only the eight low-order bits of the VPI are coded since the four high-order bits are all coded to 0 at the UNI. The VCI value field carries the VCI value assigned to the connection. The VCI identifier and the VCI value fields are included in the information element only for soft PVCCs (not for PVPCs).

8	7	6	5	4	3	2	1	Octet
colspan="8"	Calling party soft PVPC or PVCC Information element identifier	1
1 ext		Coding standard	colspan="5"	IE Instruction Field	2
colspan="8"	Length of calling party soft PVPC or PVCC contents	3
colspan="8"	Length of calling party soft PVPC or PVCC contents (continued)	4
colspan="8"	VPI identifier	5*
colspan="8"	VPI value	5.1*
colspan="8"		5.2*
colspan="8"	VCI identifier	6*
colspan="8"	VCI value	6.1*
colspan="8"		6.2*

Figure 7.42 Calling party soft PVPC or PVCC IE.

7.7.1.2 Called Party Soft PVPC or PVCC Information Element

The called party soft PVPC or PVCC IE is used to indicate the VPI (in the case of PVPC) or VPI/VCI (in the case of PVCC) values used for the PVC segment between the called connecting point and the user of a PVPC or PVCC (see Figure 7.43). These values are conveyed to the called connecting point transparently by the intermediate PNNI nodes along the connection's path.

The selection type field can take one of the three values shown in Table 7.8.

Table 7.8
The Values of the Selection Type Field

Bits								Meaning
8	7	6	5	4	3	2	1	
0	0	0	0	0	0	0	0	Any value
0	0	0	0	0	0	1	0	Required value
0	0	0	0	0	1	0	0	Assigned value

If the selection type is "any value," the called end point selects any available VPI (for PVPC) or VPI/VCI (for PVCC) at the interface. When the called end point selects the value of the connection identifier, the selected VPI/VCI value is placed in the called party soft PVPC/PVCC IE and delivered to the calling end point in the CONNECT message. In this case, the called end point codes the selection type field "assigned value."

Called party soft PVPC or PVCC Information element identifier				1
1 ext	Coding standard	IE Instruction Field		2
Length of called party soft PVPC or PVCC contents				3
Length of called party soft PVPC or PVCC contents (continued)				4
Selection type				5
VPI identifier				6*
VPI value				6.1*
				6.2*
VCI identifier				7*
VCI value				7.1*
				7.2*

Figure 7.43 Called party soft PVPC or PVCC IE.

When the type is "required value" and the indicated VPI or VPI/VCI is available, the called connecting point selects the requested value for the call. Otherwise, the connection request is rejected and a RELEASE COMPLETE message with cause "requested called party soft PVPC/PVCC not available" is sent. Accordingly, this selection type is used if it is necessary to use a predetermined connection identifier. If a failure occurs along the path of an established soft connection, the connecting end point will try to reroute the call. The new connection needs to be glued to the previous connection at the destination NI in order to preserve the end-user PVC. In this case, the calling end point will require the use of the previous connection identifier at the called end point.

7.7.2 An Overview of the Procedures

The PVPC/PVCC connecting point initiates the PVPC/PVCC establishment by sending a SETUP message. The initiation takes place when the PVPC/PVCC is initially configured, when the switching node that is the owner of the PVPC/PVCC becomes operational (e.g., power up), or after recovering from a failure. If the connection is a VP, the bearer class field of the broadband bearer capability IE is set to VP. If it is a VC, the bearer class is set to class X. To identify the VPI/VCI of the PVPC/PVCC at the destination network node, the called party soft PVPC/PVCC IE is included in the SETUP message. The called party number IE contains the configured peer PVPC/PVCC connecting point identifier while the calling party number IE contains the PVPC/PVCC connecting point's own identifier.

The SETUP message is processed within the PNNI network using the regular point-to-point and point-to-multipoint procedures. If the connection identifier between the called end point and the user is chosen by the called end point, a called party soft PVPC/PVCC IE is included in the CONNECT message and delivered to the calling end point. Upon receiving the CONNECT message, the calling endpoint node passes the connection identifier to the management entity. Negotiation of the traffic parameters is not allowed.

Connection establishment for point-to-multipoint PVCCs is initiated by the root connecting point. It is noted that point-to-multipoint VP is not supported in PNNI. When the PVCC is initially configured, a new party is added by network management. When the switching node that is the root connecting point becomes operational (e.g., power up), or during recovery from a failure, a SETUP/ADD PARTY message is sent to one of the leaf connecting points.

The setup of the first party of the point-to-multipoint PVCC is always initiated by sending a SETUP message. A bearer class of X is included in the broadband bearer capability IE. The calling and called party soft PVPC/

PVCC IEs are included in the SETUP message, the same as in point-to-point connections.

After the connection is established to the first leaf, connections can progress to additional leaves. This is achieved by sending an ADD PARTY message containing two soft PVPC/PVCC IEs for each leaf. If the ADD PARTY ACKNOWLEDGE message contains the called party soft PVPC/PVCC IE, the VPI/ VCI values of the PVCC segment between the called connecting point and the user is passed to the management entity.

8

Traffic Parameter Modification

In Chapters 2 and 4, we discussed how the traffic characteristics of a connection may be negotiated during the connection-establishment time. It is, in general, difficult to estimate the traffic requirements of an application, particularly for those that generate traffic at a varying rate (i.e., variable bit rate sources). A user may also want to increase or decrease the traffic parameters of a connection after observing for a while the service it receives from the network, particularly for visual and audio applications. Finally, the nature of the traffic may change over time. This may be the case, for example, when multiple LANs are interconnected over a wide area via an ATM backbone network.

The capability to modify the traffic characteristics of a connection after the connection is established provides flexibility for end stations to control the service they receive (or amount they pay) and to adapt to the changing traffic conditions. An alternative to modifying the connection characteristics would be to tear down the existing connection and establish a new one with the new traffic parameters. This would impose, however, a significant signaling load in the network or a discontinuity in service.

Signaling procedures used to modify a connection's traffic parameters while the connection is active are being defined by ITU-T, and they are not finalized yet. The procedures described in this section are based on the current status of these standardization efforts. All three traffic parameters—PCR, SCR, and MBS—may be modified using the modification procedures. Traffic parameter modification is applicable to only point-to-point connections with any ATM transfer capability other than ABR and ABT (it is noted that UBR service is not yet defined by ITU; hence, this capability does not apply to UBR connections at this time).

311

Modification of traffic parameters may be requested only by the end station that initiated the connection and can only be requested for connections that are in the active state (i.e., already established). That is, if a connection is being established, its traffic parameters cannot be modified. Similarly, if a connection is being cleared and a modification request is made, the clearing operation has the higher priority. This results in termination of the modification procedure (i.e., no more messages related to the modification procedure are sent across the user-network interface). This restriction was introduced to avoid modification collision.

Point-to-point connections can be unidirectional or bidirectional. The modification of traffic parameters can be requested in each direction of a bidirectional connection independent of each other. That is, modification in the forward direction may be to increase the traffic parameters, whereas it may be used to decrease the traffic parameters in the backward direction. In a given direction, all traffic parameters may be increased or decreased. In particular, it is not possible to increase a subset of the parameters while decreasing the others.

In order to modify its value after the connection is active, the corresponding parameter is required to be signaled during the connection-establishment time. For example, if the forward SCR (CLP = 0) parameter was not specified during the connection-establishment time, it cannot be modified after the connection is established. Finally, the modification request may be for all or a subset of the parameters specified during the call establishment time. In summary:

- The user who initiates the modification request expects to receive from the network a new set of traffic parameters corresponding to the modification request that are greater than or equal to the existing traffic parameters if the modification request is an increase.
- The user who initiates the modification request expects to receive from the network a new set of traffic parameters corresponding to the modification request that are less than or equal to the existing traffic parameters if the modification request is a decrease.

8.1 Modification Signaling Messages

A new set of messages are defined to support the modification of traffic parameters for point-to-point connections, as shown in Table 8.1.

8.1.1 MODIFY REQUEST

The MODIFY REQUEST message is sent by the connection owner to request modification of traffic parameters for a single connection (see Figure 8.1).

Table 8.1
Modification Signaling Messages for Point-to-Point Call Control

Message Type	Significance	Direction	Description
MODIFY REQUEST	Global	Both	Initiates a connection modification
MODIFY ACKNOWLEDGE	Global	Both	Indicates that the modification request is accepted
MODIFY REJECT	Global	Both	Indicates that the modify connection request is rejected
CONNECTION AVAILABLE	Global	Both	Indicates that the connection modification has performed in the addressed user to requesting user

Information Element	Type	Length
ATM traffic descriptor	M	8–30

Figure 8.1 Information element used in the MODIFY REQUEST message.

8.1.2 MODIFY ACKNOWLEDGE

The MODIFY ACKNOWLEDGE message is sent by the called user or network to indicate that the modify request is accepted (see Figure 8.2).

The broadband report type IE is included when the called user requires confirmation of the success of modification in the called user to calling user direction.

8.1.3 MODIFY REJECT

This message is sent by the called user or network to indicate that the modify connection request is rejected (see Figure 8.3).

Information Element	Type	Length
Broadband report type	O	4–5

Figure 8.2 Information element used in the MODIFY ACKNOWLEDGE message.

Information Element	Type	Length
Cause	M	6–34

Figure 8.3 Information element used in the MODIFY REJECT message.

8.1.4 CONNECTION AVAILABLE

This message is sent by the originating user or in response to the MODIFY ACKNOWLEDGE message that contains a broadband report type IE with a type of report field coded to "modification confirmation." It indicates the network has performed modification in the addressed user to calling user direction of transmission.

8.2 Additional Point-to-Point Call States

Two additional call states are defined to support modification, as shown in Table 8.2.

Table 8.2
Additional Call States for Point-to-Point Modification

State	Description
Modify request	A MODIFY REQUEST message has been sent to the other side of the interface
Modify received	A MODIFY REQUEST message has been received from the other side of the interface

8.3 Modification Procedures at the Requesting Entity

The requesting user can initiate modification of traffic parameters of a call/connection only if it is in the active call state. The originating user reserves corresponding resources if the requested modification is a request for an increase of one or more of the traffic parameters and additional reservation of resources is required. If the modification request is to decrease one or more ATM traffic parameter values, the end station shapes its traffic to the new parameters. In this case, the acceptance of the modify request is implicitly assumed to be accepted since a modification request to decrease the traffic parameters is always accepted by the network.

The modification request is initiated by transferring a MODIFY REQUEST message. Upon transmitting this message, the end station starts timer T360 and enters the modify request state. The MODIFY REQUEST message is required to have the same call reference value as the active connection for which the modification is requested. The inclusion of the ATM traffic descriptor IE is mandatory in the MODIFY REQUEST message. The requesting user reserves required resources if the request is to increase. If the request is to decrease, it reduces the reserved resources and starts shaping its traffic to the new parameters. These actions are taken prior to sending the MODIFY REQUEST message.

Upon receiving a MODIFY REQUEST message in the active state, the following actions are taken in the network:

- Corresponding resources are reserved if the requested modification requires additional reservation.
- The forward usage parameter control parameters are changed, if the request is to decrease one or more traffic parameters in the forward direction.
- The modification request is progressed towards the remote user.
- The network enters the modified received state.

8.3.1 Modification Acknowledgment

Upon receiving a MODIFY ACKNOWLEDGE message in the modify requested state, the user who requested the modification performs the following operations:

- Allocates corresponding resources (i.e., the connection defined by the requested ATM traffic descriptor is available for use) if the modification request was made to increase one or more traffic parameters;
- Sends a CONNECTION AVAILABLE message if the MODIFY ACKNOWLEDGE message contains a broadband report type IE indicating that confirmation of the modification request is required;
- Stops timer T360 and enters the active state.

Similarly, upon receiving a MODIFY ACKNOWLEDGE message in the modify request state, the following functions are performed in the network:

- Allocate corresponding resources if the request is for an increase, or release an appropriate amount of resources if the request was made to decrease some of the parameters.
- Change usage parameter control parameters according to the new ATM traffic parameters.
- Forward the MODIFY ACKNOWLEDGMENT message towards the initiating user.
- Stop timer T360 and enters the active state.

8.3.2 Indication of Modification Rejection

If the network or the called user rejects the modification request, the network sends a MODIFY REJECT message to the calling user at the originating UNI. Similarly, if the network determines that the modification request is not authorized or cannot be supported, it rejects the modification request. A MODIFY REJECT message includes one of the following causes:

- User cell rate not available;
- Resource unavailable, unspecified;
- Service or option not implemented, unspecified;
- Mandatory information element is missing;
- Message not compatible with call state.

If the network determines that the information received from the user is invalid (e.g., invalid ATM traffic parameter values), a MODIFY REJECT message includes one of the following cause values:

- Invalid information element contents;
- Unsupported combination of traffic parameters.

Upon receiving a MODIFY REJECT message, the originating user cancels the reservation of resources (i.e., the ATM traffic parameters are reset to their original values prior to the modification request), stops timer T360, and enters the active state.

Similarly, if a transit network receives a MODIFY REJECT message, it cancels the resources reserved when the modification request was received and reinstates the policing policy applied prior to the modification request, forwards the modification rejection to the initiating user, stops timer T360, and enters the active state.

8.3.3 No Response to Modification Request

If timer T360 expires before a response to a MODIFY REQUEST message is received, the call/connection is cleared with cause "recovery on timer expiry."

8.4 Modification Procedures at the Responding Entity

8.4.1 Modification Indication

The network indicates the arrival of a modification request at the network-to-user interface of the called user by transferring a MODIFY REQUEST message. Upon transmitting this message, it starts timer T360 and enters the modify request state. The MODIFY REQUEST message has the same call reference value as specified in the initial call/connection establishment.

Upon receiving a MODIFY REQUEST message in the active state, the called user enters the modify received state.

8.4.2 Modification Acceptance

If the user determines that the requested traffic parameters can be supported, the modification request is accepted with the following actions:

- Change the forward ATM traffic parameters as requested.
- Reduce the backward ATM traffic parameters (if any) for which decrease is requested.
- Send a MODIFY ACKNOWLEDGE message.
- Enter the active state.

If it is indicated in the MODIFICATION REQUEST message that a confirmation of the modification is required, the called end station sends a MODIFY ACKNOWLEDGE message that includes a broadband report type IE and starts timer T361 (see Figure 8.4). When modification requires an increase of backward ATM traffic parameters, it is desirable to obtain confirmation of modification before it is increased.

8.4.3 Modification Confirmation

If a CONNECTION AVAILABLE message is received by a user while timer T361 is running, the entity that received the message performs the following functions:

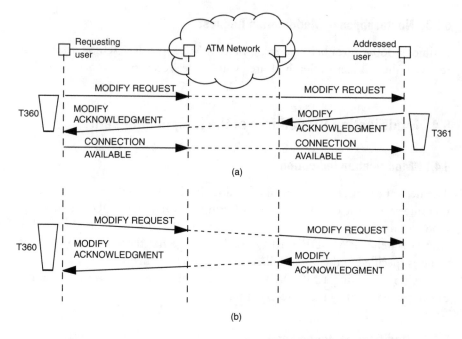

Figure 8.4 Successful modification (a) with and (b) without confirmation.

- Change the ATM traffic parameters as requested (i.e., increase backward ATM traffic parameters).
- Stop timer T361.
- Remain in the active state.

If timer T361 expires, the entity that received the message changes ATM traffic parameters as requested (i.e., increases backward ATM traffic parameters) and remains in the active state.

8.4.4 Modification Rejection

If the user is not able to support the requested increase of the ATM traffic parameters, it rejects the modify request, returns a MODIFY REJECT message with an appropriate cause value, and enters the active state (see Figure 8.5). If the responding entity is a transit network node, it also cancels the reservation of resources and reinstates the policing policy applied prior to the modification.

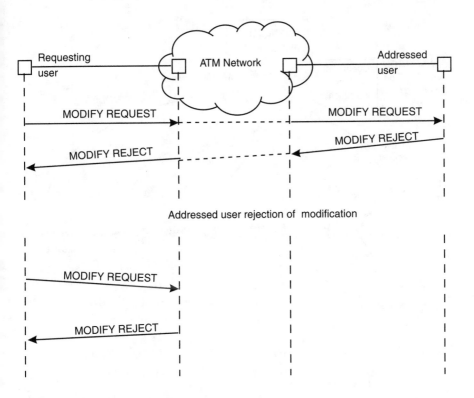

Figure 8.5 Network rejection of modification request.

8.5 Timers

Figure 8.6 illustrates the timers, connection states, messages, and primitives used in the modification procedures.

The timer specified in Table 8.3 is used at the requesting entity (i.e., the calling party at the originating UNI or the network at the destination UNI).

Similarly, Table 8.4 illustrates the timer used in the responding entity (i.e., the called party at the destination UNI or the network at the originating UNI).

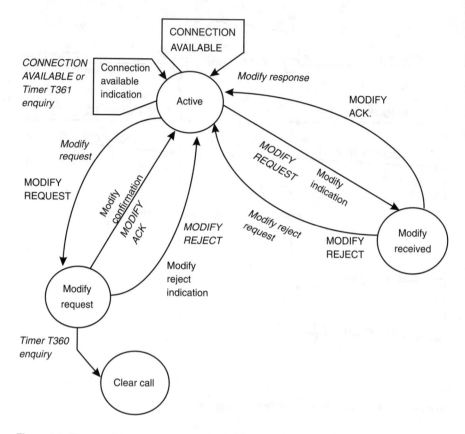

Figure 8.6 Timers, states, messages, and primitives used in modification.

Table 8.3
Timer Used in the Requesting Entity

Timer Number	Default Timeout Value	Request- ing Entity State	Cause for Start	Normal Stop	At the First Expiry	At the Second Expiry	Implementation
T360	20–30 sec	Modify requested	MODIFY REQUEST sent	MODIFY ACK. or MODIFY REJECT received	Release call	Timer is not restarted	Mandatory

Table 8.4
Timer Used in the Responding Entity

Timer Number	Default Timeout Value	Responding Entity State	Cause for Start	Normal Stop	At the First Expiry	At the Second Expiry	Implementation
T361	20 sec	Active	MODIFY ACKNOWL-EDGE sent with broadband report type indicating confirmation is required	CONNEC-TION AVAILABLE received	Commence sending in the user plane at the modified rate	Timer is not restarted	Mandatory if modification confirmation is requested (at a terminating entity)

About the Authors

Dr. Raif O. Onvural is the president of OROcom. Prior to that he was the vice president of the ATM systems development organization, Allied Telesyn International. Since starting the ATM division for Allied Telesyn International in April 1996, he has been working on the design and development of end-to-end ATM campus solutions. Prior to joining Allied Telesyn International, he was in a network architecture organization managing the network architecture department and IBM's venue owner for ATM Forum. Dr. Onvural has organized several international conferences on high-speed networks in general and ATM networks in particular. He is the coeditor of six books and the author of *ATM Networks: Performance Issues, Second Edition,* published by Artech House. He is the chairman of IEEE Computer Society Technical Committee on Computer Communications, and a member of IFIP WG 6.3 and 7.3.

Mr. Rao Cherukuri is a senior engineer at IBM, Research Triangle Park. He has made key contributions to the development of user-to-network protocols for ISDN, frame relay, and ATM. His past and current leadership positions include serving on ITU SG 11, ANSI T1, Frame Relay Forum, and ATM Forum. As a rapporteur and editor, he is responsible for the development of Q.931, Q.933, Q.2931, Q.2961, Q.2933, T1.607, T1.617, T1.618, ATM Forum UNI 3.0/3.1/4.0, and frame relay implementation agreements.

Index

AAL type 1, 48–51
    CBR rates, 50
    defined, 48
    IE format, 50
    SAR-PDU header, 48
    See also ATM adaptation layer (AAL)
AAL type 3/4, 51–52
    CPCS-SDU size, 52
    defined, 51
    IE, 52
    MID values, 51–52
AAL type 5, 52–53
    CPCS-SDU size, 53
    defined, 52
    IE format, 53
ADD PARTY ACKNOWLEDGE
            message, 124
    IEs, 124, 302
    for point-to-multipoint call/connection
            control, 301
ADD PARTY message, 122–23
    defined, 122
    IEs, 123, 302
    for point-to-multipoint call/connection
            control, 301
    receiving, 132
    sending, 131
    state diagram, 130

ADD PARTY REJECT message, 125
    IEs, 125, 303
    for point-to-multipoint call/connection
            control, 302–3
ALERTING message, 63
    adding new leaf with/without, 134
    call clearance after, 260
    IEs used in, 63, 263
    PNNI point-to-point, 263
American National Standards Institute
            (ANSI), 5, 38–39
    Committee T1, 38, 39
    standards on ATM, 39
Asynchronous transfer mode (ATM), 7
    defined, 2
    DSU, 7
    end stations, 5, 21, 22
    on-demand connections, 7
    standards, 5, 28–40
    switches, 5
ATM adaptation layer (AAL), 4
    AAL type 1, 48–51
    AAL type 3/4, 51–52
    AAL type 5, 52–53
    defined, 9
    SAP, 13
    signaling (SAAL), 9–13
    user-defined, 53

ATM adaptation layer parameter IE, 47–53
    AAL type 1, 48–51
    AAL type 3/4, 51–52
    AAL type 5, 52–53
    format, 48
    function of, 47
    indication in SETUP message, 78–82
    user-defined AAL, 53
ATM addresses, 21–23
    NSAP, 21
    private, 22
    public, 21
ATM block transfer (ABT)
        capability, *xviii*, 183–85
    ABT-DT, 185
    ABT-IT, 185
    allowable combination
        of parameters, 185, 186
    ATM traffic descriptor IE extension, 184
    defined, 183–84
    signaling, 184
    signaling procedures, 185
    See also ATM transfer capabilities
ATM Forum, 36–38
    ATC support, 165–66
    B-ICI, 7, 37
    CBR, 27
    defined, 5, 36
    element of, 36
    Enterprise Network Roundtable, 37
    interface definitions, 5
    interfaces, 6
    LAN Emulation, 38, 105, 110
    Market Awareness Committees, 37
    network management group, 38
    organization, 37
    PNNI group, 38
    SAA group, 38
    signaling group, 38
    specifications, *xviii*
    Technical Committee, 36–37
    traffic management groups, 38
    UBR, 27
    VBR, 27
    VoD, 110
ATM interfaces, 2, 5–7, 9, 10
ATM layer connection, 20–28

    QoS parameters, 24–26
    traffic parameters, 23–24
ATM networks
    design, *xvii*
    signaling structure in, 7–16
ATM traffic descriptor IE, 176, 177, 184,
        187, 272–73
    alternative, 188, 190
    minimum acceptable, 188–89
    for unrestricted/restricted digital
        information, 228
ATM transfer capabilities
        (ATCs), 26–27, 165–70
    ABR, 173–83
    ABT, 183–85
    ATM Forum support, 165–66
    ATM traffic descriptor, 169–70
    broadband bearer, 168–69
    categories, 27
    DBR, 170–72
    defined, 26
    SBR, 172–73
    signaling of, 168–70
    signaling support for, 161–91
Authority and format identifier (AFI), 21, 22
Available bit rate (ABR), *xviii*, 166
    allowable combination
        of parameters, 183
    ATM traffic descriptor IE, 176, 177
    call/connection establishment at
        destination interface, 182–83
    call/connection establishment at
        originating interface, 177–82
    cell transfer rate, 167
    flow control, 174
    information elements, 176–77
    initial cell rate (ICR), 298
    messages, 176
    parameter modification, 182
    parameters, 179
    rate decrease factor (RDF), 298
    rate increase factor (RIF), 298
    rules, 178–79
    setup parameters, 177, 178, 298
    signaling for, 176–77
    signaling procedures, 177–83

traffic parameter selection procedures
for, 297–98
transfer capability, 173–83
transient buffer exposure (TBE), 298
Available cell rate (ACR), 180, 181

Backward explicit congestion notification
(BECN) bit, 232
Bearer classes, 168–69
B-ISDN signaling, *xviii*
call/connection states, 47
ITU-T recommendations on, 32–35
B-ISDN supplementary services, 193–222
address-based, 193
address information element
relationship to, 218–19
applications using, 194
CLIP, 196–205
CLIR, 205–6
COLP, 206–14
COLR, 214–16
DDI, 194–95
definition of, 193
MSN, 195–96
SUB, 216–18
UUS, 219–22
Blocking
at a node, 292
at preceding end of link, 292–93
procedures at point of, 292
receiving clearing message
indicating, 293
at succeeding end of link, 293
Broadband bearer capability
(B-BC) IE, 226, 251
Broadband high-layer information
IE, 53–54, 226
format, 53
ISO/IEC, 53–54
Broadband integrated services digital
networks (B-ISDNs), *xvii*
cause IE, 230
interworking, 224–32
private, interworking with, 143–46
signaling flows, 228
user part (B-ISUP), 7
See also B-ISDN supplementary services

Broadband intercarrier interface (B-ICI), 7
Broadband low-layer information
IE, 54–55, 226
format, 54
multiple, 55
in SETUP message, 55
user information protocol fields, 55
Broadband repeat indicator IE, 61
Burst tolerance (BT), 161, 162
Busy state, 136

Call-clearing messages, 83–85
RELEASE, 84
RELEASE COMPLETE, 84–85
Call/connection clearing, 83–87
clear collision, 87
leaf-initiated joins, 159
messages, 83–85
procedures, 85–87
state diagrams, 87, 88
Call/connection-establishment
procedures, 69–83
AAL parameters indication in SETUP
message, 78–82
connection establishment at the
originating interface, 69–74
at destination interface, 74, 74–78, 78
end-to-end transit delay IE, 82–83
PNNI point-to-point, 258–62
state diagrams, 78, 81, 82
Call control messages, 62–68
ALERTING, 63
CALL PROCEEDING, 63
CONNECT, 64
CONNECT ACKNOWLEDGE, 64
list of, 62
PNNI point-to-point, 263–68
SETUP, 64–66
Called party number IE, 55–56
addressing/numbering combinations, 56
defined, 55
format, 55
Called party subaddress IE, 56
Calling line identification presentation
(CLIP), 196–204
actions at destination switch, 200–202
actions at originating switch, 197–200

Calling line identification presentation
    (CLIP) (continued)
  defined, 196
  destination switch processing, 202
  interworking, 202–3
  originating side processing, 198
  procedures, 203–4
  two-calling party number IE
      delivery, 203–4
  uses, 196–97
  See also B-ISDN supplementary services
Calling line identification restriction
    (CLIR), 199, 205–6
  defined, 205
  interworking, 205–6
  originating network side procedure, 206
  precedence, 206
  See also B-ISDN supplementary services
Calling party number IE, 56–57
  defined, 56
  extension bit, 67
  format, 57
Calling party subaddress IE, 57
CALL PROCEEDING message, 63
  call termination after, 258–59
  IEs used in, 63, 264
  PNNI point-to-point, 263–64
Call reference error, 94–95
Calls/connections
  acceptance, 71–72, 77–78
  B-ISDN connection states, 47
  confirmation indication, 71
  at destination interface, 74–78
  establishment state diagrams, 78, 81, 82
  incoming/outgoing, 42
  invalid control information, 70–71
  proceeding, 71
  rejection, 72–73
  request, 69–70
  states, 45, 46
Call state IE, 100
Capability set 1 (CS-1), 34
Capability set 2 (CS-2), 34
Cause IE, 57–58
  B-ISDN, 230
  defined, 57
  format, 58

N-ISDN, 230
Cell delay variation (CDV), 24
  maximum, 24
  one-point, 24–25
  peak-to-peak, 24, 25
  two-point, 25
Cell delay variation tolerance
    (CDVT), 24, 161
  defined, 162
  at private UNI, 24
Cell error ratio (CER), 24
Cell loss priority (CLP), 162–63
Cell loss ratio (CLR), 24, 25
Cell transfer delay (CTD), 24
Clear collision, 87
Code sets
  active, 68
  shifting procedures, 68
Compatibility checking, 74–76
  called side terminal equipment, 75–76
  called user side, 75
  information categories, 74–75
CONNECT ACKNOWLEDGE
    message, 64
Connected line identification presentation
    (COLP), 206–14
  actions at destination switch, 209–11
  actions at originating local
      exchange, 208–9
  defined, 206–7
  information elements, 207–8
  information when special arrangement
      applies, 213
  information when special arrangement
      does not apply, 212
  interworking, 211–14
  processing at destination switch, 214
  processing at originating switch, 215
  See also B-ISDN supplementary services
Connected line identification restriction
    (COLR), 214–16
  defined, 214–15
  interworking, 216
  processing, 217
  See also B-ISDN supplementary services
Connected number IE, 207
Connected subaddress IE, 207–8

Connection characteristics
negotiation, 187–91
acceptance, 191
alternative traffic parameter, 190
destination user, 189
IEs, 188
minimum acceptable ATM traffic
parameter, 188–89
Connection establishment, 69–74
call/connection acceptance, 71–72
call/connection confirmation
indication, 71
call/connection proceeding, 71
call/connection rejection, 72–73
call/connection request, 69–70
at destination interface - network side, 79
at destination interface - user side, 80
invalid call/connection control
information, 70–71
QoS and traffic parameter selection
procedures, 70
state diagram - network side, 81
state diagram - user side, 82
transit network selection, 73–74
Connection identifier allocation, 268–72
switched virtual channels, 269–70
switched virtual paths, 270–72
Connection identifier IE, 58–59
defined, 58
format, 58
VP-associated signaling field, 59
VPCI, 58–59
Connections. *See* Calls/connections
Connection scope IE, 268
CONNECT message, 64
IEs used in, 64, 265
PNNI point-to-point, 263–64
Constant bit rate (CBR), *xviii*
ATM Forum, 27
service class, 166
CPCS-SDU
AAL type 3/4, 52
AAL type 5, 53
forward and backward maximum
values, 80
maximum size negotiation, 78–81
Crankback, 287–95

causes, 289
defined, 287
IE, 258, 288–89, 303
IE updated topology state
parameters, 294–95
level, 288
to next higher level, 294
procedures for crankback level and
blocked transit, 291–94
procedures for point-to-multipoint calls/
connections, 304–5
procedure summary, 295, 296
request, 290–91

Data exchange interface (DXI), 7
Data-link connection identifier (DLCI), 232
Data service unit (DSU), 7
Designated transit list (DTL), 280–87
IE, 280–82
originator, 284
postprocessing procedures, 287
preprocessing, 283
processing, 282–86
processing errors, 291
processing summary, 286–87
terminator, 285–86
See also Private NNI (PNNI)
Deterministic bit rate (DBR) transfer
capability, 170–72
allowable combinations, 171–72
network commitment, 171
uses, 170–71
See also ATM transfer capabilities
(ATCs)
Direct dialing in (DDI), 194–96
defined, 194
support, 195
See also B-ISDN supplementary services
Discard eligibility (DE) bit, 233
Domain-specific part (DSP), 21
DROP PARTY ACKNOWLEDGE
message, 125–26
IEs, 126, 304
for point-to-multipoint call/connection
control, 303–4
DROP PARTY message, 125
IEs in, 125, 303

DROP PARTY message (continued)
for point-to-multipoint call/connection
control, 303

Endpoint reference IEs, 119, 126–27, 256
errors, 140
illustrated, 120
Endpoint state IEs, 119, 127
End-to-end transit delay IE, 59–60, 278–80
defined, 59
format, 59
handling, 82–83, 147–50
maximum value, 60
point-to-multipoint signaling
and, 147–50
point-to-point signaling and, 82–83
subfield identifiers, 67
Error conditions, 93–99, 139–43
call reference, 94–95
DTL processing, 291
endpoint reference IE-related, 140
general information-element, 95
leaf-initiated joins, 160
mandatory information
element, 96, 141–42
mandatory information element
content, 160
mandatory information element
missing, 160
message sequence, 95, 140–41
message too short, 94
message type, 95, 140–41
nonmandatory information-
element, 97–98
protocol discriminator, 94
reachability, 290
resource, 291
status enquiry procedure, 142–43
Error procedures, 103–5
information-element, 104–5
messages with insufficient
information, 105
unexpected/unrecognized message
type, 103–4
European Telecommunication Standards
Institute (ETSI), 5, 40

Forward error correction (FEC), 49
Forward explicit congestion notification
(FECN) bit, 232
Frame discard, 185–87
Frame relay (FR), 223, 232–38
to ATM interworking, 234
ATM terminal with, 234
defined, 232
emulated, 233
emulation IEs, 235
header formats, 233
interworking, 237–38
mapping of call setup messages, 235–36
traffic parameters mapping, 236–37

General IE errors, 95
Generic call admission control (GCAC), 289

Idle state, 136
Information elements (IEs), 19–20
AAL type 1, 50
in ADD PARTY ACKNOWLEDGE
message, 124, 302
in ADD PARTY message, 123, 302
in ADD PARTY REJECT
message, 125, 303
address, supplementary services
and, 218–19
in ALERTING message, 63, 263
ATM adaptation layer parameter, 47–53
ATM traffic descriptor, 176, 177, 184,
187, 228, 272–73
broadband bearer capability
(B-BC), 128, 226, 251
broadband high-layer information
(B-LLI), 53, 226
broadband low-layer information
(B-HLI), 54–55, 226
broadband repeat indicator, 61
called party number, 55–56
called party subaddress, 56
calling party number, 56–57
calling party subaddress, 57
in CALL PROCEEDING
message, 63, 264
call state, 100
cause, 57–58, 230

code set extensions, 68
coding rules, 66–68
COLP service, 207–8
connected number, 207
connected subaddress, 207–8
connection identifier, 58–59
connection scope, 268
in CONNECT message, 64, 265
crankback, 258, 288–89, 303
in DROP PARTY ACKNOWLEDGE
    message, 126, 304
in DROP PARTY message, 125, 303
DTL, 280–82
for emulating frame relay, 235
endpoint reference, 119, 120,
    126–27, 256
endpoint state, 119, 127
end-to-end transit delay, 59–60
extensions of octets in, 67
format, 19
generic transport identifier, 113
IE action indicator, 19, 20
IE identifier, 19
IE instruction, 19
information subfield identifier
    structure, 66–68
in LEAF SETUP FAILURE
    message, 151
in LEAF SETUP REQUEST
    message, 152
LIJ call identifier IE, 152–53
LIJ leaf sequence number IE, 153
LIJ parameters IE, 153
message length, 94
in MODIFY ACKNOWLEDGE
    message, 313–14
in MODIFY REJECT message, 314
in MODIFY REQUEST message, 313
N-BC, 225
N-HLC, 225
N-LLC, 225
notification indicator, 263
in PARTY ALERTING
    message, 124, 303
progress indicator, 231
PVPC/PVCC, 307–9
Q.2931 call setup messages, 47–62

quality of service parameter, 60–61
RESTART ACKNOWLEDGE
    message, 90
restart indicator, 89–90
RESTART message, 89
in SETUP message, 65, 266
shift, 68
in STATUS message, 99
transit network selection, 61–62, 268
UUS, 220
Initial domain identifier (IDI), 21–22
Initial domain part (IDP), 21
Integrated services digital network
    (ISDN), *xviii*
    broadband (B-ISDN), *xvii*, 143–46
    circuit mode, 224–32
    narrowband (N-ISDN), 223
International Telecommunications Union
    (ITU-T), *xviii*, 28, 29
    ATM call model, 28
    defined, 29
    interface definitions, 5
    study groups, 29–30
    subworking parties, 30
    working parties, 29–30
    See also ITU-T Recommendations
Interworking, 223–38
    B-ISDN to N-ISDN direction, 229
    CLIP, 202–3
    CLIR, 205–6
    COLP, 211–14
    COLR, 216
    examples, 225
    frame relay, 237–38
    ISDN circuit mode, 224–32
    N-ISDN to B-ISDN direction, 228–29
    SUB, 218
    UUS, 222
Interworking function (IWF), 228–29
    mapping function, 229–30
    N-ISDN to B-ISDN direction, 228–29
    at the user plane, 234
IP over ATM, 113–16
    defined, 113
    model characteristics, 113
    Q.2931 signaling IEs, 116
ISDN circuit mode, 224–32

ISDN circuit mode (continued)
  interworking between N-ISDN and
    B-ISDN, 228–30
  notification of interworking, 231
  tones and announcements, 231–32
ITU Radiocommunication Sector
  (ITU-R), 28, 29
ITU Telecommunication Development
  Sector (ITU-D), 29
ITU-T Recommendations, 31–32
  I.150, 249
  I.365, 233
  I.555, 233
  on B-ISDN signaling, 32–35
  on supplementary services, 35–36
  Q.931, 225, 226
  Q.933, 233, 235, 238
  Q.2100, 33
  Q.2110, 33
  Q.2120, 33
  Q.2130, 33
  Q.2931, *xviii*, 33, 42, 194, 225,
    233, 249, 258
  Q.2932, 33
  Q.2933, 233, 235, 238
  Q.2951, 33–34, 193, 203
  Q.2957, 34, 193, 219
  Q.2959, 34
  Q.2961, 34
  Q.2962, 34
  Q.2963, 35
  Q.2964, 35
  Q.2971, *xviii*, 35, 117, 249

LAN Emulation, 105–10
  ATM Forum, 105, 110, 118
  UNI (LUNI), 6
Leaf-initiated join (LIJ), 150–60
  to active network LIJ call, 154–56
  call/connection, 159
  capability, 150–60
  error condition handling, 160
  as first party to point-to-multipoint
    call, 156–58
  IEs, 152–53
  LEAF SETUP FAILURE message, 151

LEAF SETUP REQUEST
  message, 151–52
  LIJ call identifier IE, 152–53
  LIJ leaf sequence number IE, 153
  LIJ parameters IE, 153
  network call root creation, 153–54
  party clearing, 159
  as root LIJ call, 158–59
  timers, 160
LEAF SETUP FAILURE message, 151
LEAF SETUP REQUEST message, 151–52
  defined, 151
  IEs, 152
Link state update (LSU) messages, 244
Local-area networks (LANs), 1

Mandatory IE errors, 96
Mapping
  of call setup messages, 235–36
  IWF function, 229–30
  traffic parameters, 236–37
Message type/message sequence errors, 95
Minimum cell rate (MCR), 167, 173
MODIFY REJECT message, 313–14
MODIFY REQUEST message, 312–13
Multiconnection calls, 28
Multiple subscriber number (MSN), 195–96
  defined, 195
  private ISDN stations and, 196
  provided by calling user, 195–96
  user's view, 195
  See also B-ISDN supplementary services
Multiplexing identification field
  AAL type 3/4 values, 51–52
  range negotiation, 81–82

Narrowband bearer capability (N-BC)
  IE, 225
Narrowband high-layer compatibility
  (N-HLC) IE, 225
Narrowband ISDN (N-ISDN), 223
  backbone, 224
  cause IE, 230
  circuit-mode, 224
  development of, 223
  interworking, 224–32
  N-BC IE for, 226

signaling flows, 228
Narrowband low-layer compatibility
    (N-LLC) IE, 225
Network LIJ calls, 153–56
    active, leaf joins to, 154–56
    inactive, leaf joins to, 157
    root creation of, 153–54
    See also Leaf-initiated join
Network service access point (NSAP), *xviii*
    addresses, 21
    address structure, 22
Network-to-network interface (NNI), 5, 6
    private, 6, 7, 239–310
    public, 6, 7
Nonmandatory IE errors, 97–98
Non-real-time VBR (NRT-VBR), *xviii*
    application traffic characteristics, 166
    service guarantees and, 167
Notification indicator IE, 263

Packet-switched networks, 1–2
    telecommunications networks vs., 2
    virtual-circuit-oriented, 1–2
PARTY ALERTING message, 124
    IEs, 124, 303
    for point-to-multipoint call/connection
        control, 301–2
Party clearing, 159
Party dropping, 135–39, 146
    leaf-initiated, 137
    network-initiated, 138
    root-initiated, 137, 138
    state transitions, 139
Party states, 121–22
    defined, 121
    list of, 121
    See also UNI point-to-multipoint
        signaling
Peak cell rate (PCR), 23, 161, 311
    cell multiplexing effect, 23
    defined, 161
    example, 161
Permanent virtual channel connection
        (PVCC). *See* Soft PVPC/PVCC
Permanent virtual path connection (PVPC).
        See Soft PVPC/PVCC
PNNI point-to-multipoint

adding a party, 300
call setup, 256–57, 298–305
call states, 253
crankback procedures, 304–5
first party setup, 299–300
messages, 251, 301–4
setup illustration, 299
See also Private NNI (PNNI)
PNNI point-to-point
    call clearance, 260–61
    call connection/control
        procedures, 257–68
    call/connection establishment, 258–62
    call setup, 253–56
    call states, 252–53
    message format, 257–58
    messages, 250
    state diagram, 252
    successful call/connection setup, 262
    See also Private NNI (PNNI)
PNNI point-to-point call control
        messages, 263–68
    ALERTING, 263
    CALL PROCEEDING, 263–64
    CONNECT, 264
    SETUP, 264–68
Point-to-multipoint connections, 20
    defined, 117
    illustrated, 117
    setup, 119
    using, 118–19
    See also UNI point-to-multipoint
        signaling
Point-to-multipoint messages
    ADD PARTY, 122–23
    ADD PARTY ACKNOWLEDGE, 124
    ADD PARTY REJECT, 125
    DROP PARTY, 125
    DROP PARTY ACKNOWLEDGE,
        125–26
    list of, 120–21
    overview, 119–21
    PARTY ALERTING, 124
    PNNI, 251
    See also UNI point-to-multipoint
        signaling

Point-to-multipoint signaling
        procedures, 127–39
    adding party, 129–31
    adding party establishment, 133–35
    adding party rejection, 131–32
    connection setup to first party, 128–31
    party dropping, 135–39
Point-to-point connections, 20
Primitives, 13–16, 42
    confirm, 13
    exchange of, 14
    indication, 13
    list of, 14
    request, 13
    response, 13
    SAP, 15
    See also Signaling AAL (SAAL)
Private B-ISDN interworking, 143–46
    incoming add party request, 144
    party dropping, 146
    response to add party request, 145–46
Private NNI (PNNI), 239–310
    call control, 248
    connection identifier allocation, 268–72
    connection setup, 253–57
    control plane, 248
    crankback, 287–95
    defined, 239
    designated transit list, 280–87
    DTLs, 244–47
    framework overview, 240–49
    GCAC, 289
    hierarchical model, 241–42
    hierarchical peer groups, 243
    links, 241, 247
    negotiation procedures, 272
    network illustrations, 240, 242
    nodes, 282
    point-to-multipoint call setup, 256–57
    point-to-multipoint call states, 253
    point-to-point call setup, 254–56
    point-to-point call states, 251–53
    protocol control, 248
    QoS parameter selection
        procedures, 275–80
    reference model, 247
    restart message procedure, 297

    restart procedure, 295–97
    routing, 240, 243, 244
    SAAL layer, 248
    signaling, 244, 247, 249–53
    signaling channels across, 254–56
    signaling messages, 249–51
    signaling procedures, 268–98
    soft PVPC/PVCC, 305–10
    specifications, 239–40
    traffic parameter negotiation, 272–75
    traffic parameter selection procedures
        for ABR connections, 297–98
Private UNI, 7, 24
Progress indicator IE, 231
Protocol discriminator error, 94

Q.2931 Recommendation, xviii, 33, 42,
        194, 258
    call states, 100
    global interface states, 101
    IEs for ISDN circuit-mode services, 226
    IEs in IP over ATM, 116
    IEs in LAN Emulation service, 111–12
    primitives between call control to
        signaling entity, 44
    primitives between SAAL and, 43
    primitives between signaling entity to call
        control, 44
    signaling IEs, 114–15
    See also ITU-T Recommendations
Q.2971 Recommendation, xviii,
        35, 117, 249
Quality of service (QoS)
    classes, 163–64, 165
    parameters, 24–26
    parameter selection procedures, 275–80
    parameters used in ATM, 164
    preceding side parameter processing, 276
    requirements in PNNI networks, 275
    succeeding side parameter
        processing, 278
Quality of service parameter IE, 60–61
    classes, 60–61
    format, 60
    procedures, 70

Real-time VBR (RT-VBR), xviii, 165, 166

RELEASE COMPLETE message, 84–85
RELEASE message, 84
Remote operations service element
     (ROSE), 33
Resource reservation protocol (RSVP), 2
RESTART ACKNOWLEDGE
     message, 89–90
  defined, 89
  IEs, 90
Restart indicator IE, 89–90
  defined, 89
  format, 90
RESTART message, 87–89
  IEs, 89
  receiving, 91–93, 94
  sending, 90–91
Restart procedure, 87–93, 139
  information elements, 87–90
  PNNI, 295–97
  point-to-multipoint, 139
  point-to-point, 87–93
  RESTART ACKNOWLEDGE
     message, 89–90
  RESTART message, 87–89, 90–93

Service access point (SAP), 13
Service-specific connection-oriented
     protocol (SSCOP), 9
  connection establishment, 15
  connection release, 16
  defined, 11
  functions, 11
  PDUs, 12
  signals, 13
Service-specific coordination
     function (SSCF), 9
  defined, 9–11
  services, 12
  signals, 13
SETUP message, 64–66
  AAL parameters indication in, 78–82
  IEs in, 65, 266
  mapping, 235–36
  no reply to first, 258
  response to, 76–77
  sending, 64
Signaling

for ABR, 176–77
for ABT, 184
associated, 269
channels, 7, 8
defined, 1
nonassociated, 270
PNNI, 239–310
purposes, 3–4
support for ATM transfer
     capabilities, 161–91
traffic parameters used in, 163
UNI point-to-multipoint, 117–60
UNI point-to-point, 41–116
use of, 1
Signaling AAL (SAAL), 9–13
  components, 9–11
  connection reset, 98–99
  CPCS, 9
  PNNI, 248
  primitives, 13–16
  primitives between Q.2931 and, 43
  SAP, 13
  SSCF, 9–11, 12–13
  SSCS, 9
  SSOP, 9, 11–12
Signaling message format, 17–19
  call reference field, 17–18
  illustrated, 17
  length field, 19
  message type field, 18
  PNNI, 257–58
  type field, 18
Signaling messages, 4
  ABR, 176
  call-clearing, 83–85
  call/control control, 62–68
  information elements (IEs), 19
  PNNI, 249–51
  point-to-multipoint control, 122–26
  significance levels, 19
  transfer, 16
  UUS, 220
Signaling virtual channel (SVC), 9
  multiple interfaces controlled by, 10
  single interface controlled by, 9
Soft PVPC/PVCC, 305–10
  called party IE, 308–9

Soft PVPC/PVCC (continued)
  calling party IE, 307
  concept, 306
  connecting point, 309
  establishing/releasing, 305
  establishment procedures, 306
  IEs, 307–9
  parameters, 307
  procedure overview, 309–10
State diagrams
  ADD PARTY message, 130
  call/connection clearing, 87, 88
  calls/connection
        establishment, 78, 81, 82
  PNNI point-to-point, 252
Statistical bit rate (SBR) transfer
        capability, 172–73
  defined, 172
  NRT-SBR, 175
  RT-SBR, 174
  signaling procedures for, 172–73
  uses, 172
  See also ATM transfer capabilities
        (ATCs)
Status enquiry, 99–103
  information elements, 99–100
  procedure, 100–101, 142–43
  STATUS ENQUIRY message, 99
  STATUS message, 99, 101–3
STATUS ENQUIRY message, 99
STATUS message, 99, 101–3
  IEs, 99
  receiving, 101–3, 143
  sending, 143
Structured data transfer (SDT) block size, 51
Subaddressing (SUB), 216–18
  defined, 216
  functions, 216–17
  interworking, 218
  normal operation, 218
  without priority arrangement, 217
  See also B-ISDN supplementary services
Subfield identifiers, 66–68
  example of, 67
  extension mechanism, 67–68
  using, 66–67
Sustainable cell rate (SCR), 24, 161, 311

defined, 162
using, 162
Switched multimegabit data service
        (SMDS), 7
Switched virtual channels, 269–70
  associated signaling, 269
  nonassociated signaling, 270
Switched virtual paths, 270–72
Synchronous residual time stamp (SRTS)
        method, 49

Tagging, 162
Timers, 105, 106–9, 147–49
  leaf-initiated joins, 160
  in network side, 106–7, 149
  point-to-multipoint signaling, 147–49
  point-to-point signaling, 106–9
  T301 processing, 260
  T303 processing, 259
  T308 expiry, 261
  T310 processing, 259
  traffic parameter modification, 319–21
  in user side, 108–9, 148, 160
Traffic parameter modification, 311–21
  acceptance, 317
  acknowledgment, 315–16
  confirmation, 317–18
  indication, 317
  no response to request, 317
  procedures at requesting entity, 314–17
  procedures at responding entity, 317–19
  rejection, 318–19
  rejection indication, 316
  signaling messages, 312–14
  successful, 318
  timers, 319–21
Traffic parameters, 23–24
  additional point-to-point states, 314
  alternative, 190
  mapping, 236–37
  minimum acceptable, 188–89
  negotiation, 272–75
  QoS and, 70
  selection procedures for ABR, 297–98
  used in signaling, 163
Transit network selection IE, 61–62
  defined, 61

format, 61
in PNNI point-to-point, 268

UNI point-to-multipoint signaling, 117–60
control messages, 122–26
end-to-end transit delay IE
handling, 147–50
error condition handling, 139–43
framework overview, 119–22
interworking with private
B-ISDNs, 143–46
leaf-initiated join capability, 150–60
message overview, 119–21
party states, 121–22
procedures, 127–39
restart procedure, 139
timers, 147, 148, 149
UNI point-to-point signaling, 41–116
AAL connection reset, 98–99
B-ISDN call/connection states, 47
call/connection clearing, 83–87
call/connection control, 42–47
call/connection control messages, 62–68
call/connection-establishment
procedures, 69–83
call/connection states, 45, 46
error condition handling, 93–98
error procedures, 103–5
examples, 105–16
information elements, 47–62
restart procedure, 87–93
status enquiry, 99–103
timers, 105, 106–9
Unspecified bit rate (UBR), *xviii*
ATM Forum, 27
service class, 166–67

User-defined AAL, 53
User-to-network Interface (UNI)
LAN Emulation (LUNI), 6
private, 7, 24
signaling specifications, 41
specification, 5, 6
See also UNI point-to-multipoint
signaling; UNI point-to-point
signaling
User-to-user signaling (UUS), 219–22
communication, 219
defined, 219
information elements, 220
interworking, 222
messages, 220
services, 220
UNI signaling procedures, 221–22
See also B-ISDN supplementary services

Variable bit rate (VBR)
ATM Forum, 27
NRT-VBR, *xviii*, 166
RT-VBR, 166
Video on demand (VoD), 110–13
ATM Forum specifications, 110
ATM virtual connection for, 110–13
defined, 110
Q.2931 signaling IEs, 114–15
Virtual circuit identifier (VCI), 7
any, 269, 270
exclusive, 269, 270
Virtual path connection identifier (VPCI), 8
Virtual path identifier (VPI), 7
exclusive, 269, 270
values, 8

The Artech House Telecommunications Library

Vinton G. Cerf, Series Editor

Access Networks: Technology and V5 Interfacing, Alex Gillespie

Advanced High-Frequency Radio Communications, Eric E. Johnson, Robert I. Desourdis, Jr., et al.

Advanced Technology for Road Transport: IVHS and ATT, Ian Catling, editor

Advances in Computer Systems Security, Vol. 3, Rein Turn, editor

Advances in Telecommunications Networks, William S. Lee and Derrick C. Brown

Advances in Transport Network Technologies: Photonics Networks, ATM, and SDH, Ken-ichi Sato

An Introduction to International Telecommunications Law, Charles H. Kennedy and M. Veronica Pastor

Asynchronous Transfer Mode Networks: Performance Issues, Second Edition, Raif O. Onvural

ATM Switches, Edwin R. Coover

ATM Switching Systems, Thomas M. Chen and Stephen S. Liu

Broadband: Business Services, Technologies, and Strategic Impact, David Wright

Broadband Network Analysis and Design, Daniel Minoli

Broadband Telecommunications Technology, Byeong Lee, Minho Kang and Jonghee Lee

Cellular Mobile Systems Engineering, Saleh Faruque

Cellular Radio: Analog and Digital Systems, Asha Mehrotra

Cellular Radio: Performance Engineering, Asha Mehrotra

Cellular Radio Systems, D. M. Balston and R. C. V. Macario, editors

CDMA for Wireless Personal Communications, Ramjee Prasad

Client/Server Computing: Architecture, Applications, and Distributed Systems Management, Bruce Elbert and Bobby Martyna

Communication and Computing for Distributed Multimedia Systems, Guojun Lu

Communications Technology Guide for Business, Richard Downey, et al.

Community Networks: Lessons from Blacksburg, Virginia, Andrew Cohill and Andrea Kavanaugh, editors

Computer Networks: Architecture, Protocols, and Software, John Y. Hsu

Computer Mediated Communications: Multimedia Applications, Rob Walters

Computer Telephone Integration, Rob Walters

Convolutional Coding: Fundamentals and Applications, Charles Lee

Corporate Networks: The Strategic Use of Telecommunications, Thomas Valovic

The Definitive Guide to Business Resumption Planning, Leo A. Wrobel

Digital Beamforming in Wireless Communications, John Litva and Titus Kwok-Yeung Lo

Digital Cellular Radio, George Calhoun

Digital Hardware Testing: Transistor-Level Fault Modeling and Testing, Rochit Rajsuman, editor

Digital Switching Control Architectures, Giuseppe Fantauzzi

Digital Video Communications, Martyn J. Riley and Iain E. G. Richardson

Distributed Multimedia Through Broadband Communications Services, Daniel Minoli and Robert Keinath

Distance Learning Technology and Applications, Daniel Minoli

EDI Security, Control, and Audit, Albert J. Marcella and Sally Chen

Electronic Mail, Jacob Palme

Enterprise Networking: Fractional T1 to SONET, Frame Relay to BISDN, Daniel Minoli

Expert Systems Applications in Integrated Network Management, E. C. Ericson, L. T. Ericson, and D. Minoli, editors

FAX: Digital Facsimile Technology and Applications, Second Edition,
Dennis Bodson, Kenneth McConnell, and Richard Schaphorst

FDDI and FDDI-II: Architecture, Protocols, and Performance,
Bernhard Albert and Anura P. Jayasumana

Fiber Network Service Survivability, Tsong-Ho Wu

Future Codes: Essays in Advanced Computer Technology and the Law,
Curtis E. A. Karnow

Guide to Telecommunications Transmission Systems,
Anton A. Huurdeman

A Guide to the TCP/IP Protocol Suite, Floyd Wilder

Implementing EDI, Mike Hendry

Implementing X.400 and X.500: The PP and QUIPU Systems, Steve Kille

Inbound Call Centers: Design, Implementation, and Management,
Robert A. Gable

Information Superhighways Revisited: The Economics of Multimedia,
Bruce Egan

Integrated Broadband Networks, Amit Bhargava

International Telecommunications Management, Bruce R. Elbert

International Telecommunication Standards Organizations,
Andrew Macpherson

Internetworking LANs: Operation, Design, and Management,
Robert Davidson and Nathan Muller

Introduction to Document Image Processing Techniques,
Ronald G. Matteson

Introduction to Error-Correcting Codes, Michael Purser

An Introduction to GSM, Siegmund Redl, Matthias K. Weber and
Malcom W. Oliphant

*Introduction to Radio Propagation for Fixed and Mobile
Communications,* John Doble

Introduction to Satellite Communication, Bruce R. Elbert

Introduction to T1/T3 Networking, Regis J. (Bud) Bates

Introduction to Telephones and Telephone Systems, Second Edition,
    A. Michael Noll

Introduction to X.400, Cemil Betanov

LAN, ATM, and LAN Emulation Technologies, Daniel Minoli and
    Anthony Alles

Land-Mobile Radio System Engineering, Garry C. Hess

LAN/WAN Optimization Techniques, Harrell Van Norman

LANs to WANs: Network Management in the 1990s, Nathan J. Muller
    and Robert P. Davidson

*Minimum Risk Strategy for Acquiring Communications Equipment and
    Services,* Nathan J. Muller

Mobile Antenna Systems Handbook, Kyohei Fujimoto and J. R. James,
    editors

*Mobile Communications in the U.S. and Europe: Regulation,
    Technology, and Markets,* Michael Paetsch

Mobile Data Communications Systems, Peter Wong and David Britland

Mobile Information Systems, John Walker

Networking Strategies for Information Technology, Bruce Elbert

Packet Switching Evolution from Narrowband to Broadband ISDN,
    M. Smouts

Packet Video: Modeling and Signal Processing, Naohisa Ohta

Performance Evaluation of Communication Networks,
    Gary N. Higginbottom

Personal Communication Networks: Practical Implementation,
    Alan Hadden

Personal Communication Systems and Technologies, John Gardiner and
    Barry West, editors

Practical Computer Network Security, Mike Hendry

Principles of Secure Communication Systems, Second Edition,
    Don J. Torrieri

Principles of Signaling for Cell Relay and Frame Relay, Daniel Minoli
    and George Dobrowski

Principles of Signals and Systems: Deterministic Signals, B. Picinbono

Private Telecommunication Networks, Bruce Elbert

Radio-Relay Systems, Anton A. Huurdeman

RF and Microwave Circuit Design for Wireless Communications,
  Lawrence E. Larson

The Satellite Communication Applications Handbook, Bruce R. Elbert

Secure Data Networking, Michael Purser

Service Management in Computing and Telecommunications,
  Richard Hallows

Signaling in ATM Networks, Raif O. Onvural, Rao Cherukuri

Smart Cards, José Manuel Otón and José Luis Zoreda

Smart Card Security and Applications, Mike Hendry

Smart Highways, Smart Cars, Richard Whelan

Successful Business Strategies Using Telecommunications Services,
  Martin F. Bartholomew

Super-High-Definition Images: Beyond HDTV, Naohisa Ohta, et al.

Telecommunications Deregulation, James Shaw

Television Technology: Fundamentals and Future Prospects,
  A. Michael Noll

Telecommunications Technology Handbook, Daniel Minoli

Telecommuting, Osman Eldib and Daniel Minoli

Telemetry Systems Design, Frank Carden

Teletraffic Technologies in ATM Networks, Hiroshi Saito

*Toll-Free Services: A Complete Guide to Design, Implementation, and
  Management,* Robert A. Gable

Transmission Networking: SONET and the SDH, Mike Sexton and
  Andy Reid

Troposcatter Radio Links, G. Roda

Understanding Emerging Network Services, Pricing, and Regulation,
  Leo A. Wrobel and Eddie M. Pope

Understanding GPS: Principles and Applications, Elliot D. Kaplan, editor

Understanding Networking Technology: Concepts, Terms and Trends, Mark Norris

UNIX Internetworking, Second Edition, Uday O. Pabrai

Videoconferencing and Videotelephony: Technology and Standards, Richard Schaphorst

Voice Recognition, Richard L. Klevans and Robert D. Rodman

Wireless Access and the Local Telephone Network, George Calhoun

Wireless Communications in Developing Countries: Cellular and Satellite Systems, Rachael E. Schwartz

Wireless Communications for Intelligent Transportation Systems, Scott D. Elliot and Daniel J. Dailey

Wireless Data Networking, Nathan J. Muller

Wireless LAN Systems, A. Santamaría and F. J. López-Hernández

Wireless: The Revolution in Personal Telecommunications, Ira Brodsky

Writing Disaster Recovery Plans for Telecommunications Networks and LANs, Leo A. Wrobel

X Window System User's Guide, Uday O. Pabrai

For further information on these and other Artech House titles, including previously considered out-of-print books now available through our In-Print-Forever™ (IPF™) program, contact:

Artech House
685 Canton Street
Norwood, MA 02062
781-769-9750
Fax: 781-769-6334
Telex: 951-659
email: artech@artech-house.com

Artech House
Portland House, Stag Place
London SW1E 5XA England
+44 (0) 171-973-8077
Fax: +44 (0) 171-630-0166
Telex: 951-659
email: artech-uk@artech-house.com

Find us on the World Wide Web at:
www.artech-house.com